WHAT STAR?

To my grandsons James, Thomas and Christopher –
future students of the stars.

WHAT STAR?

BRIAN JONES

CHARTWELL
BOOKS

This edition published in 2015 by
CHARTWELL BOOKS
an imprint of The Quarto Group
142 West 36th Street, 4th Floor,
New York, NY 10018 USA
www.QuartoKnows.com

Copyright © 2015
Regency House Publishing Limited
The Manor House, High Street
Buntingford, Hertfordshire
SG9 9AB, United Kingdom

For all editorial enquiries, please contact:
www.regencyhousepublishing.com

ISBN-13: 978-0-7858-3324-6

Printed in China

10 9 8 7 6 5 4 3 2

The author and publishers have made every effort to
ensure that the information contained in this book
was accurate at the time of going to press.

WARNING
Never look at the sun with the naked eye or any
stargazing equipment.

Page 2: Looking towards the center of our Galaxy.

Page 3: The Horsehead Nebula, a dark nebula in Orion.

*Right: Orion graces a starlit sky, as seen from Glencoe in
the Highlands of Scotland.*

*Page 8-9: The Milky Way spans the sky above the
Israeli desert.*

Contents

LIST OF CONSTELLATIONS

CONSTELLATION	MEANING / GENITIVE FORM	ORDER OF SIZE	BRIGHTEST STAR
Andromeda	Andromeda/Andromedae	19	Sirrah
Antlia	The Air Pump/Antliae	62	Alpha (α) Antliae
Apus	The Bird of Paradise/Apodis	67	Alpha (α) Apodis
Aquarius	The Water Carrier/Aquarii	10	Sadalsuud
Aquila	The Eagle/Aquilae	22	Altair
Ara	The Altar/Arae	63	Alpha (α) and Beta (β) Arae
Aries	The Ram/Arietis	39	Hamal
Auriga	The Charioteer/Aurigae	21	Capella
Boötes	The Herdsman /Boötis	13	Arcturus
Caelum	The Graving Tool/Caeli	81	Alpha (α) Caeli
Camelopardalis	The Giraffe/Camelopardalis	18	Beta (β) Camelopardalis
Cancer	The Crab/Cancri	31	Al Tarf
Canes Venatici	The Hunting Dogs/Canum Venaticorum	38	Cor Caroli
Canis Major	The Great Dog/Canis Majoris	43	Sirius
Canis Minor	The Little Dog/Canis Minoris	71	Procyon
Capricornus	The Sea Goat/Capricorni	40	Deneb Algiedi
Carina	The Keel/Carinae	34	Canopus
Cassiopeia	Cassiopeia/Cassiopeiae	25	Shedar
Centaurus	The Centaur/Centauri	9	Alpha (α) Centauri
Cepheus	Cepheus/Cephei	27	Alderamin
Cetus	The Whale/Ceti	4	Deneb Kaitos
Chamaeleon	The Chameleon/Chamaeleontis	79	Alpha (α) Chamaeleontis
Circinus	The Compasses/Circini	85	Alpha (α) Circini
Columba	The Dove/Columbae	54	Phakt
Coma Berenices	Berenice's Hair/Comae Berenicis	42	Beta (β) Comae Berenicis
Corona Australis	The Southern Crown /Coronae Australis	80	Alfecca Meridiana
Corona Borealis	The Northern Crown/Coronae Borealis	73	Alphecca
Corvus	The Crow/Corvi	70	Gienah
Crater	The Cup/Crateris	53	Delta (δ) Crateris
Crux	The Cross/Crucis	88	Acrux
Cygnus	The Swan/Cygni	16	Deneb
Delphinus	The Dolphin/Delphini	69	Rotanev
Dorado	The Goldfish/Doradus	72	Alpha (α) Doradus
Draco	The Dragon/Draconis	8	Etamin
Equuleus	The Little Horse/Equulei	87	Kitalpha
Eridanus	The River/Eridani	6	Achernar
Fornax	The Furnace/Fornacis	41	Alpha (α) Fornacis
Gemini	The Twins/Geminorum	30	Pollux
Grus	The Crane/Gruis	45	Alnair
Hercules	Hercules/Herculis	5	Kornephoros
Horologium	The Pendulum Clock/Horologii	58	Alpha (α) Horologii
Hydra	The Water Snake/Hydrae	1	Alphard
Hydrus	The Little Water Snake/Hydri	61	Beta (β) Hydri

WHAT STAR?

Indus	The Indian/Indi	49	Alpha (α) Indi
Lacerta	The Lizard/Lacertae	68	Alpha (α) Lacertae
Leo	The Lion/Leonis	12	Regulus
Leo Minor	The Little Lion/Leonis Minoris	64	Praecipua
Lepus	The Hare/Leporis	51	Arneb
Libra	The Scales/Librae	29	Zubeneschamali
Lupus	The Wolf/Lupi	46	Alpha (α) Lupi
Lynx	The Lynx/Lyncis	28	Alpha (α) Lyncis
Lyra	The Lyre/Lyrae	52	Vega
Mensa	The Table Mountain/Mensae	75	Alpha (α) Mensae
Microscopium	The Microscope/Microscopii	66	Gamma (γ) Microscopii
Monoceros	The Unicorn/Monocerotis	35	Beta (β) Monocerotis
Musca	The Fly/Muscae	77	Alpha (α) Muscae
Norma	The Level/Normae	74	Gamma2 (γ2) Normae
Octans	The Octant/Octantis	50	Nu (ν) Octantis
Ophiuchus	The Serpent Bearer/Ophiuchi	11	Ras Alhague
Orion	Orion/Orionis	26	Rigel
Pavo	The Peacock/Pavonis	44	Peacock
Pegasus	The Winged Horse/Pegasi	7	Enif
Perseus	Perseus/Persei	24	Algenib
Phoenix	The Phoenix/Phoenicis	37	Ankaa
Pictor	The Painter's Easel/Pictoris	59	Alpha (α) Pictoris
Pisces	The Fishes/Piscium	14	Al'farg
Piscis Austrinus	The Southern Fish/Piscis Austrini	60	Fomalhaut
Puppis	The Poop or Stern/Puppis	20	Naos
Pyxis	The Mariner's Compass/Pyxidis	65	Alpha (α) Pyxidis
Reticulum	The Net/Reticuli	82	Alpha (α) Reticuli
Sagitta	The Arrow/Sagittae	86	Gamma (γ) Sagittae
Sagittarius	The Archer/Sagittarii	15	Kaus Australis
Scorpius	The Scorpion/Scorpii	33	Antares
Sculptor	The Sculptor/Sculptoris	36	Alpha (α) Sculptoris
Scutum	The Shield/Scuti	84	Alpha (α) Scuti
Serpens	The Serpent/Serpentis	23	Unukalhai
Sextans	The Sextant/Sextantis	47	Alpha (α) Sextantis
Taurus	The Bull/Tauri	17	Aldebaran
Telescopium	The Telescope/Telescopii	57	Alpha (α) Telescopii
Triangulum	The Triangle/Trianguli	78	Beta (β) Trianguli
Triangulum Australe	The Southern Triangle/Trianguli Australis	83	Atria
Tucana	The Toucan/Tucanae	48	Alpha (α) Tucanae
Ursa Major	The Great Bear/Ursae Majoris	3	Alioth
Ursa Minor	The Little Bear/Ursae Minoris	56	Polaris
Vela	The Sail/Velorum	32	Gamma (γ) Velorum
Virgo	The Virgin/Virginis	2	Spica
Volans	The Flying Fish/Volantis	76	Beta (β) Volantis
Vulpecula	The Fox/Vulpeculae	55	Anser

INTRODUCTION

Although the night sky is a fascinating place, and has held the attention of astronomers and stargazers for thousands of years, your first impressions when looking up into the heavens can be those of confusion. Even a casual glance at the sky, particularly on a really clear, dark night, can reveal hundreds or even thousands of pinpoints of light. However, initial bewilderment soon gives way to a semblance of order in that some stars are obviously much brighter than others, and that many of the stars appear to be arranged in distinctive patterns. These patterns of stars are known as constellations. The sky is divided into a total of 88 constellations, some fairly large and covering relatively expansive areas of sky with others being quite small and often seemingly there simply to fill up gaps between the larger star groups. Whether they are large or small, the constellations are the key to your exploration of the heavens, the location and exploration of one constellation at a time ensuring that you slowly but surely get to know your way around, and start to feel at home under the star-filled sky.

The main section of this book is devoted to the constellations, with each star group having its own individual entry. These include a chart showing the main outline of the constellation together with mythology attached to the character or object that it depicts. Many of the star groups were formed thousands of years ago by the ancient Greeks who thought they could see the shapes of animals and other characters in them. It is often difficult to envisage how they identified the character the constellation was supposed

HOW TO SEE THE STARS

Given clear, dark skies, the naked eye will reveal the constellations, the Milky Way and objects such as the Andromeda Galaxy.

Binoculars are an ideal instrument for the beginner and will bring out the colors of many stars as well as revealing a number of star clusters, nebulae and galaxies.

Once you have acquainted yourself with the night sky, a telescope is often the next step for the backyard astronomer.

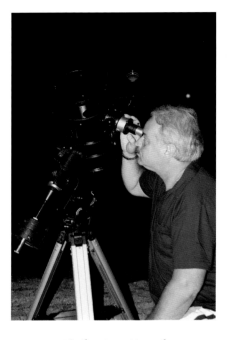

to represent in the star patterns they devised, although the names they gave to the star groups are still in use today.

Many other constellations have been devised and added to star charts in more recent times though the more modern additions generally have no mythology associated with them. In these cases the information provided in this book includes details of the astronomer who devised the group, when the constellation was drawn up and the

ABOVE: Expensive telescopes are generally much more powerful and will allow you to pick out objects in more detail.

object or character that the astronomer in question intended the constellation to depict.

Prior to the 1920s there was no well-defined list of constellations, with numerous astronomers throughout the ages having concocted new star groups in an attempt to leave their mark in the heavens. It was clear that something had to be done to tidy up the mess and in the

1920s the International Astronomical Union formally adopted the 88 constellations that we use today. Boundaries were drawn up for each one, the whole process ensuring that the officially-accepted constellations neatly covered the entire heavens. Constellations are still the accepted system of identifying the locations of objects in various parts of the night sky.

Although details are given as to the visibility of each constellation from different locations on the Earth, it must be borne in mind that, when viewed from locations at or near the limits stated, the

ABOVE: A beautiful star-filled sky plays host to the brilliant Venus, seen here hanging low over the horizon.

constellations in question may be situated fairly low down in the sky. In these cases, whether you can make them out at all will depend a great deal on the absence of any cloud, mist or light pollution over the horizon above which the constellations are situated.

ABOVE: Professional astronomers have access to very large telescopes such as those housed at Kitt Peak National Observatory, seen here perched high in the Quinlan Mountains in Arizona, USA.

OPPOSITE: During this exposure, a tripod-mounted unguided camera was trained on the night sky. The camera shutter was then left open to produce the star trails seen here, which are the result of the rotation of the Earth and the subsequent motion of the stars through the sky.

Identification of many of the fainter constellations depicted in this book is facilitated by the inclusion on the chart of a nearby prominent guide star. Should you have any problems identifying a particular constellation, reference to the seasonal and circumpolar star charts will enable you to pick out the individual star patterns within their general area and in relation to neighboring groups. You should be able to track down and identify most of the constellations without too much trouble, although some of the fainter groups may present a challenge in that they are composed of faint stars and are difficult to pick out against the background star fields. Antlia (the Air Pump), Camelopardalis (the Giraffe), Lacerta (the Lizard), Lynx (the Lynx) and Vulpecula (the Fox) definitely fall into this

category, although in these cases a little patience will produce its rewards.

In addition, information is provided on the main stars forming the constellation, including the name or designation of the star (*see Glossary – Star Names*), its brightness (*see Glossary – Magnitudes*), the distance to the star (*see Glossary – Light Year*) and, in many cases, the star's color (*see Glossary – Star Colors*).

Unlike the astronomers of thousands of years ago, we now know that the constellations are not necessarily associations or groups of stars, but in the main are formed from stars that only happen to lie more or less in the same line of sight as seen from Earth. For example, if we were able to view a constellation, such as Cassiopeia or Triangulum Australe,

from locations elsewhere in space, the stars that we see forming these groups would be arranged quite differently from the way we see them from our vantage point on Earth.

Other information given includes details of interesting objects within the constellation, including double and variable stars, star clusters, nebulae and galaxies. In many cases, only a selection of the main objects in each constellation is included, all objects described (including the main stars forming the constellation) being depicted on the charts.

As you spend time under the night sky, you will soon realize that not only do the stars we see seem to rise in the east, cross the sky and set in the west, but also that we see completely different stars and constellations at different times of the year. To get a full understanding of why this happens we need to examine the reasons behind it. So, before we commence our journey around the sky, let's take a look at some of the basics.

THE CHANGING SKY

The overall positions of the stars and constellations we see in the sky remain fixed relative to each other. However, the

ABOVE: The beautiful Milky Way straddles the sky, as imaged from the Elqui Valley in Chile.

PAGES 16-17: This view of the northern sky shows the constellation Ursa Minor, its leading star Polaris (the Pole Star) visible just to the right of centre. Part of the Plough (in Ursa Major) can be seen to its left with Alcor and Mizar, the prominent naked-eye double star in the Plough 'handle', visible near the top of this image.

general view we get of the heavens is constantly changing, this being down to the movements of our planet rather than any motion of the stars themselves. For example, the Earth rotates on its axis from west to east, and if we could view the Earth from a point in space above the northern hemisphere, we would see our planet rotating in a counterclockwise direction. The axial rotation of our planet produces night and day, daytime occurring on the side of the Earth which happens to be facing the Sun and night taking place when that side of our planet moves around and is facing away from the Sun.

The consequences of the Earth's rotation from west to east are not limited simply to the creation of night and day. From the point of view of anyone standing on the Earth's surface, the axial rotation of our planet causes the stars and other objects in the sky to appear to travel in the opposite direction, rising in the east and crossing the sky before setting in the west, presenting an ever-changing vista for the backyard astronomer.

STARS FOR ALL SEASONS

The seasonal star charts included in this book show the main stars and constellations visible during the different seasons. There are charts for Northern Summer/Southern Winter; Northern Autumn/Southern Spring; Northern Winter/Southern Summer; and Northern Spring/Southern Autumn. In addition, there are charts showing the main circumpolar stars, these being the constellations lying around the north and south celestial poles (see Glossary – Celestial Poles and Circumpolar Star) and which (as far as these particular charts are concerned) never set as seen from mid-northern or mid-southern latitudes.

The rotation of the Earth around its axis, as described above, is one of the two main motions which affect our planet, the other being its orbital motion around the Sun. As we have seen, the nightly movement of the stars across the sky is caused by the rotation of the Earth on its axis. On the other hand, the net result of the Earth's orbital motion around the Sun is that we can see different regions of the celestial sphere at different parts of the year. In other words, there is a change in the appearance of the sky during different seasons.

The reason that we see different stars during different seasons is that, throughout the Earth's annual journey around the Sun, our view of the celestial sphere alters slightly from night to night. At any particular point in its orbit, or during any particular season, we can only see the stars that are in the opposite direction in the sky to the Sun. In other words, the stars that we can observe are visible from the side of the Earth which is facing away from the Sun at the time, and which is therefore experiencing night.

During the night time we are looking out on that section of the celestial sphere opposite that in which the Sun lies. Of course, there are stars in the sky during daylight hours, although these can't be seen due to the fact that their light is swamped by the glare of the Sun. To put it another way, we can only see the stars during the night!

As the Earth travels along its orbit, the section of sky presented to us alters very slightly from night to night. As the days and seasons progress, the stars and constellations on view change, and a different section of the celestial sphere is presented to us in winter to that which we see in spring, summer and autumn. For example, during northern winter/southern summer, the night sky plays host to the majestic constellation Orion, although six months later the Earth will be at the opposite point in its orbit around the Sun and the stars of Orion will be in the sky during the daytime and therefore invisible to us. However, constellations such as Cygnus, Lyra and Aquila, which were lost to view six months earlier, will be seen gracing the night sky. The upshot of this is that, over the course of a year (during which time the Earth will complete a full orbit of the Sun), we are able to see all the stars and constellations that are visible from our latitude on Earth. The stars and constellations have not actually moved, of course. It is just that we are looking at them from different points in the Earth's orbit around the Sun, so they appear in the sky at different times.

As well as showing the stars and constellations visible during any particular season, the seasonal star charts provide a useful backup to help you identify individual constellations as they depict each group in relation to those around it. Also, because they overlap slightly with the seasonal chart to either side and to the circumpolar charts, you should have little trouble linking the different sections of the celestial sphere together.

Hopefully the journey you will be taking around the heavens will be a rewarding one, and will introduce you to many of the wonders the night sky holds. Happy stargazing!

THE NORTH CIRCUMPOLAR SKY

A circumpolar star or constellation is one that never sets as seen from a given location, but always remains above the observer's horizon and can be viewed all year round. This chart depicts stars which are circumpolar as seen from mid-northern latitudes and which can be seen to be centred on Polaris, the brightest star in the constellation Ursa Minor (the Little Bear). Also known as the Pole Star, Polaris lies very close to the north celestial pole and is always seen to lie due north.

Observers at mid-northern latitudes will see the conspicuous pattern of stars forming the Plough, together with the rest of Ursa Major (the Great Bear), close to the zenith or overhead point during mid-evenings in spring. The Great Bear's place is taken over by Draco (the Dragon) during the summer months whilst Cassiopeia wheels overhead during autumn. In winter, their place is taken by the indistinct and rambling Camelopardalis (the Giraffe). This faint group stretches from the region of sky immediately adjoining Polaris southwards towards an area close to Capella, the brightest star in the constellation Auriga (the Charioteer) and Algenib, the leading star in Perseus. The trio of bright guide stars Polaris, Capella and Algenib, coupled with clear, dark skies, will be essential in helping you to pick out this obscure constellation.

The Plough is an excellent guide to finding your way around the north circumpolar sky. By holding the chart above your head and rotating it to match up the pattern of stars shown on the chart with their locations in the sky, you will be able to identify the stars and constellations you can actually see. Once you've picked out the main constellations shown here, and providing the sky is dark and clear, the rest of the (fainter) stars on this chart can be identified.

Formed from the seven brightest members of Ursa Major, the conspicuous pattern of stars known as the Plough is the most prominent and well-known group in this region. It is also an ideal starting point for finding your way around this area of sky, including the location of the Pole Star. A line extended from Merak through Dubhe in the Plough will lead you to Polaris, from where the rest of the stars in Ursa Minor can be picked out.

The Plough acts as a useful direction finder to many other stars and constellations including the distinctive W- or M-shaped Cassiopeia. Continuing the line from Merak and Dubhe roughly as far again past Polaris will bring you to Caph, the star marking the end of the pattern formed from the five brightest stars in Cassiopeia. According to legend, Cassiopeia was the wife of King Cepheus of Ethiopia, the constellation depicting her husband seen here adjoining Cassiopeia.

Other groups in this area of sky include the long and winding Draco (the Dragon) which curls its way around Ursa Minor and which is formed from stars which are generally a little brighter than those in the faint constellation Lynx (the Lynx). The stars forming Lynx are all quite faint and, as a consequence, clear dark skies are essential in order to pick out the zigzag line of stars which make up this dim group. As neither Camelopardalis or Lynx contain any particularly bright stars you may need of a pair of binoculars to help you identify them.

BELOW: Hanging in the twilight sky we see the Plough, actually a part of the much larger constellation Ursa Major.

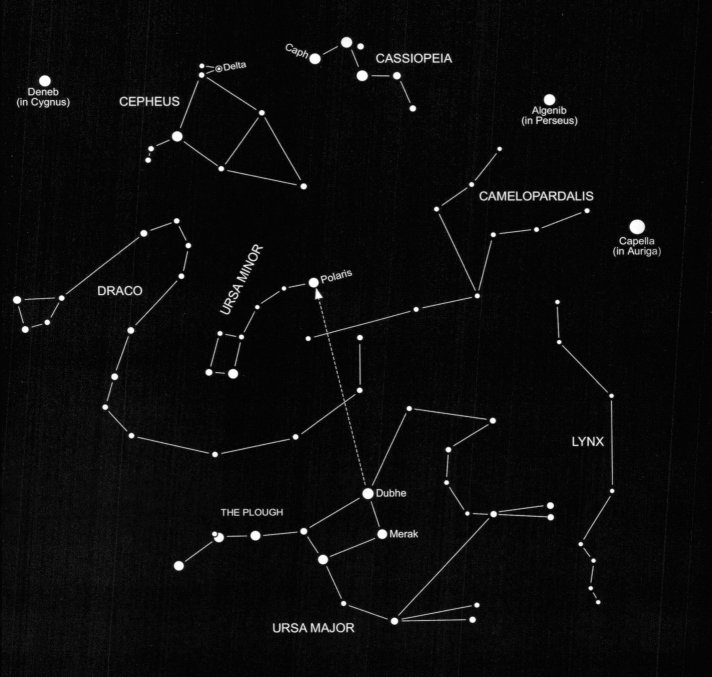

Backyard astronomers at mid-northern latitudes will see the brilliant star Capella in Auriga located at or around the overhead point at this time of year. This region of sky contains a number of prominent stars, Capella being the leading member of the constellation Auriga (the Charioteer). This distinctive group takes the form of a large circlet of stars which, at first sight, appears to include El Nath, a star which is actually a member of the neighboring constellation Taurus (the Bull).

To the southeast of Auriga we find Gemini (the Twins), its two brightest stars Castor and Pollux particularly prominent, as is Procyon, the leading star in Canis Minor (the Little Dog) situated a little to the south of Pollux. The conspicuous Perseus (Perseus) borders Auriga to the west, with the straggling line of faint stars forming Lynx (the Lynx) lying to the northeast of Gemini and Auriga. The faint constellation Cancer (the Crab) adjoins Gemini and Canis Minor to the east with the tiny circlet of stars forming the Head of Hydra (the Water Snake) visible immediately to its south.

By far the most prominent constellation on view at this time is Orion (Orion), the distinctive quadrilateral formed from its brightest stars unmistakable and almost impossible to miss. Orion is visible in its entirety from almost every inhabited part of the world with Betelgeuse and Rigel particularly prominent. Following the line formed by the three stars in the Belt of Orion to the northwest takes us first of all to Aldebaran, the leading star in Taurus, extending the line further bringing us to

the Pleiades, the well-known open star cluster in the northern reaches of the celestial bull. Following the same line of stars towards the southeast leads us to Sirius, the brightest star in Canis Major (the Great Dog) with the small group of stars forming the constellation Lepus (the Hare) visible immediately to the south of Orion.

To the east of Orion we find the faint constellation Monoceros (the Unicorn) occupying the region of sky between the two celestial dogs Canis Major and Canis Minor. Eridanus (the River) meanders away from a point just to the northwest of Rigel at the foot of Orion, the northern extremity of Eridanus marked by the star Cursa. Flowing southwards to a point deep inside the southern sky, Eridanus passes to the west of the faint constellations Caelum (the Graving Tool) and Horologium (the Pendulum Clock) before reaching its southernmost point, marked by the bright star Achernar.

Several fainter groups can be found by using Achernar as a guide including Reticulum (the Net), located to the east of Achernar and Horologium. The tiny constellation Caelum is bordered by the distinctive shape of Columba (the Dove) to the east with Dorado (the Goldfish) and Pictor (the Painter's Easel) a little to the south of these. Both Dorado and Pictor are situated immediately to the west of the brilliant star Canopus in the constellation Carina (the Keel).

Puppis (the Poop or Stern) and Vela (the Sail) adjoin Carina to the north, two of the stars in Carina and two in Vela together making up the prominent asterism known as the 'False Cross'. The trio of

constellations Carina, Puppis and Vela were devised by the French astronomer Nicolas Louis de Lacaille, being formed from the stars making up the old Argo Navis (the Ship Argo). Believing Argo Navis to be too large and unwieldy, Lacaille divided the celestial ship into the three separate constellations that appear on modern star charts. The nautical theme of this region of sky is supported by the tiny constellation Pyxis (the Mariner's Compass) which lies immediately to the east of Puppis and is another of Lacaille's creations.

ABOVE: The constellation Orion, with the Orion Nebula (M42) prominent to the south of the three stars forming the Belt of Orion.

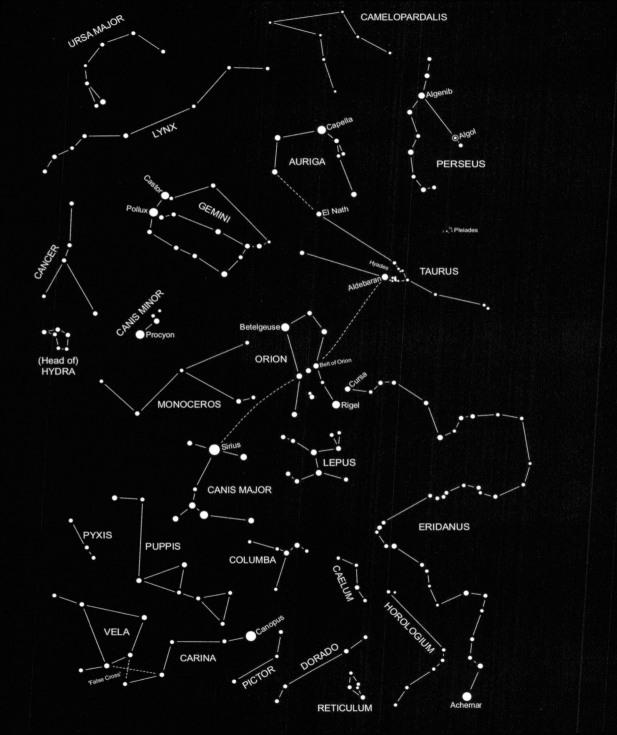

THE NORTHERN SPRING / SOUTHERN AUTUMN SKY

The conspicuous outline of the Plough, formed from the seven brightest stars in the constellation Ursa Major (the Great Bear), is a familiar sight for observers at mid-northern latitudes, being located at or near the overhead point during spring evenings. The rest of the Ursa Major can be seen extending out to the west of the Plough, while immediately to its north is part of the long and winding constellation Draco (the Dragon). Merak and Dubhe, both located at the end of the 'bowl' of the Plough, act as pointers to Polaris, the Pole Star (as shown on the chart for The North Circumpolar Sky). Polaris is the brightest star in Ursa Minor (the Little Bear) and marks the location of the north celestial pole.

Extending the curve formed by the stars in the Plough handle southwards will take you to Arcturus in the constellation Boötes (the Herdsman), continuing the line further to the south eventually leading you to Virgo (the Virgin) and its brightest star Spica. The area of sky to the west of this curve plays host to the two faint constellations Canes Venatici (the Hunting Dogs) and Coma Berenices (Berenice's Hair).

Other dim groups that occupy this region include Leo Minor (the Little Lion) and Lynx (the Lynx), both of which lie immediately south of the paws of the Great Bear. Far more prominent is Leo (the Lion) which can be found south of Leo Minor and to the northwest of Virgo. Although dark and clear skies will be needed in order to pick out Leo Minor and Lynx, the distinctive shape of Leo is far more conspicuous. However, clear skies will be needed to identify the small and faint constellation Sextans (the Sextant) located just to the south of the bright star Regulus in Leo.

Slightly more prominent than Sextans are Corvus (the Crow) and Crater (the Cup), both of which lie to the southeast of Sextans and immediately to the north of the long and winding constellation Hydra (the Water Snake). Once you locate Alfard, the brightest star in Hydra, the rest of the celestial water snake can be seen to stretch away eastwards and can be picked out with the help of binoculars. Traveling along Hydra will eventually lead you to the Mira-type variable star R Hydrae, located a little to the south of Spica in Virgo.

The two tiny and inconspicuous constellations Antlia (the Air Pump) and Pyxis (the Mariner's Compass) occupy an area of sky to the south of Hydra. To the south of these is the larger and brighter Vela (the Sail) which in turn is bordered by Carina (the Keel). A combination of four stars straddling the border between Vela and Carina form the asterism known as the 'False Cross' which is often confused with Crux (the Cross), the distinctive form of which is located some way to the east. Crux itself is surrounded on three sides by the southern reaches of Centaurus (the Centaur) with Alpha and Beta Centauri, the two leading stars of Centaurus, particularly prominent. The constellation Lupus (the Wolf) borders Centaurus to the east.

BELOW: The ruddy glow of the red giant Gamma (γ) Crucis is prominent in this view which shows the main outline of the tiny constellation Crux at upper right of image with the two nearby stars Alpha (α) and Beta (β) Centauri towards the lower left.

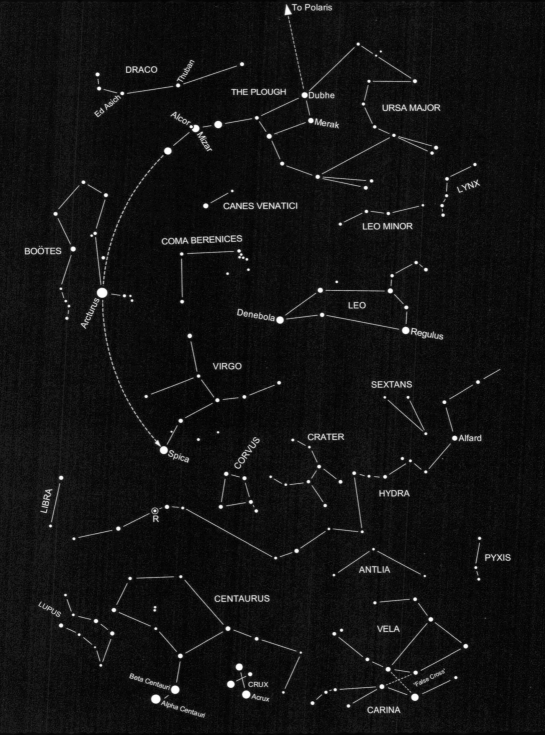

THE NORTHERN SUMMER / SOUTHERN WINTER SKY

The skies of northern summer are dominated by the three constellations Cygnus (the Swan), Lyra (the Lyre) and Aquila (the Eagle). Situated high in the evening sky as seen from mid-northern latitudes, the triangle formed from the bright stars Deneb in Cygnus, Vega in Lyra and Altair in Aquila is a familiar sight to backyard astronomers in the northern hemisphere. Known as the Summer Triangle, this trio of stars is unmistakable and from here many of the other constellations on the chart can be located.

The trio of diminutive constellations Vulpecula (the Fox), Sagitta (the Arrow) and Delphinus (the Dolphin) are seen immediately to the east of the Summer Triangle while to the north of Lyra is the small but fairly prominent group of stars forming the head of Draco (the Dragon). A little to the west of Lyra is the distinctive quadrilateral of stars which marks the central regions of the constellation Hercules. Moving further west we see the circlet of stars forming Corona Borealis (the Northern Crown), just beyond which is part of Boötes (the Herdsman), the whole of Boötes being depicted on the chart showing the Northern Spring / Southern Autumn sky.

Adjoining Aquila to the southwest is the faint constellation Scutum (the Shield), to the northwest of which is the line of faint stars forming Serpens Cauda (the 'tail' of the Serpent). Making your way along Serpens Cauda will bring you to Ophiuchus (the Serpent Bearer), to the west of which is another line of stars forming Serpens Caput (the 'head' of the Serpent). Greek mythology identifies Ophiuchus as Asclepius, the god of medicine, holding the head of the serpent in his left hand and the tail in his right, thereby splitting the constellation Serpens into two parts. Scutum, Serpens and Ophiuchus should be visible to the naked eye if the sky is fairly dark and clear, although a pair of binoculars will help you to pick them out.

Scorpius (the Scorpion) lies southwest of Ophiuchus, the reddish glow of its brightest star Antares unmistakable. The name of this star translates as 'rival of Mars' and alludes to the fact that when Mars (often referred to as the red planet) and Antares are in the same area of sky, the two objects rival each other for prominence.

Scorpius is flanked to the west by Libra (the Scales) and to the east by the large and sprawling Sagittarius (the Archer). The circlet of stars forming Corona Australis (the Southern Crown) can be seen immediately to the south of Sagittarius with the faint constellation Microscopium (the Microscope) found a short way to the east. The region of sky to the south of this group is occupied by the two faint constellations Indus (the Indian) and Telescopium (the Telescope).

Although clear, dark skies may be needed in order to pick out Microscopium, Indus and Telescopium, the tiny but brighter Ara (the Altar), located just to the south of the prominent curve of stars forming the Sting of Scorpius, will be easier to identify. Finally, the prominent pair of stars Alpha and Beta Centauri are useful guides to locating the three faint constellations Norma (the Level), Lupus (the Wolf) and Circinus (the Compasses).

BELOW: This image shows the constellation Scorpius straddling the Milky Way, with the reddish glow of its brightest star Antares clearly seen just to the upper right of center. The curve of stars forming the Sting of Scorpius lie at the bottom edge of the image.

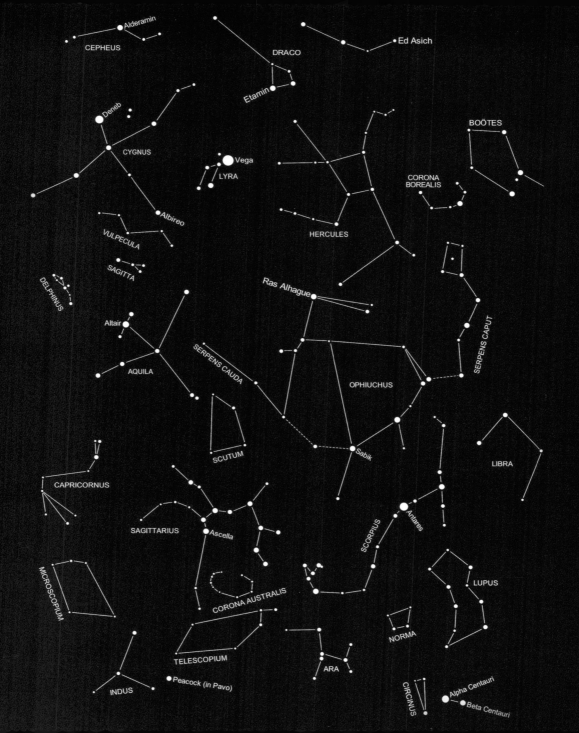

THE NORTHERN AUTUMN / SOUTHERN SPRING SKY

Observers at mid-northern latitudes will find the distinctive form of Cassiopeia (Cassiopeia) at or near the overhead point during autumn evenings. Her husband Cepheus (Cepheus) can be seen immediately to her west, the southern regions of Cepheus, including the variable star Delta Cephei, depicted here. The zigzag line of faint stars that form the tiny constellation Lacerta (the Lizard) border Cepheus to the south, this faint and somewhat obscure group probably difficult to locate without optical aid. This is in contrast to the bright star Deneb in the neighboring constellation Cygnus (the Swan). Located to the west of Lacerta, Deneb is one of the stars forming the famous asterism known as the Summer Triangle (*see Cygnus*) which extends beyond the borders of this chart but is depicted in full on the chart showing the Northern Summer / Southern Winter sky.

The prominent line of stars forming Andromeda (Andromeda) can be found to the south of Cassiopeia. The westernmost star in Andromeda is Sirrah, which also marks the northeastern corner of the adjoining and very conspicuous Square of Pegasus (the Winged Horse), this huge quadrilateral of stars being a striking feature of the night sky at this time of year. Pegasus is one of the largest constellations in the entire sky, the rest of the group extending westward towards its brightest star Enif, adjoining which we find the tiny constellation Equuleus (the Little Horse).

Two tiny but conspicuous groups lie to the southeast of Andromeda, these being Triangulum (the Triangle) and Aries (the Ram). Just to the northeast of Andromeda is the constellation Perseus (Perseus), its bright star Algenib prominent and, just to the south of Algenib, the famous variable star Algol. The large but faint Pisces (the Fishes) meanders from a point west of Triangulum to the area immediately south of the Square of Pegasus.

Aquarius (the Water Carrier) borders Pegasus to the southwest and can be found by extending a line from Scheat southwards through Markab, both in the Square of Pegasus. This will take you to the bright star Fomalhaut in Piscis Austrinus (the Southern Fish) although before you reach Fomalhaut the line will pass immediately east of the star Skat in Aquarius from where you should be able to make out the rest of this large and sprawling group. Unless the sky is dark and clear you may need the help of binoculars to pick out the stars of Aquarius, as well as those which make up the constellation Capricornus (the Goat), which can be found immediately to its southwest.

Bordering the southeastern edge of Pisces is Cetus (the Whale), another of the faint constellations that grace this region of sky. The fairly prominent star Deneb Kaitos depicts the Whale's tail, this star located slightly east of a line extended southwards from Sirrah through Algenib (both in the Square of Pegasus). Cetus also plays host to the famous long-period variable star Mira, to the southeast of which we see a section of the long and winding Eridanus (the River).

If the sky is dark and clear the fainter stars in Piscis Austrinus, as well as the neighboring obscure constellations Sculptor (the Sculptor) and Fornax (the Furnace) should be visible, both of which lie to the east of Fomalhaut.

The line of stars forming Grus (the Crane) borders Piscis Austrinus to the south, Grus itself being encircled by the faint constellations Microscopium (the Microscope), Indus (the Indian), Tucana (the Toucan) and Phoenix (the Phoenix). Ankaa, the brightest star in Phoenix, can be picked up fairly easily to the southeast of Fomalhaut, the southern reaches of Eridanus (the River), together with its leading star Achernar, located nearby. Achernar is an excellent guide to help you to locate many of the fainter constellations in this region of the sky, including the nearby trail of faint stars forming Horologium (the Pendulum Clock).

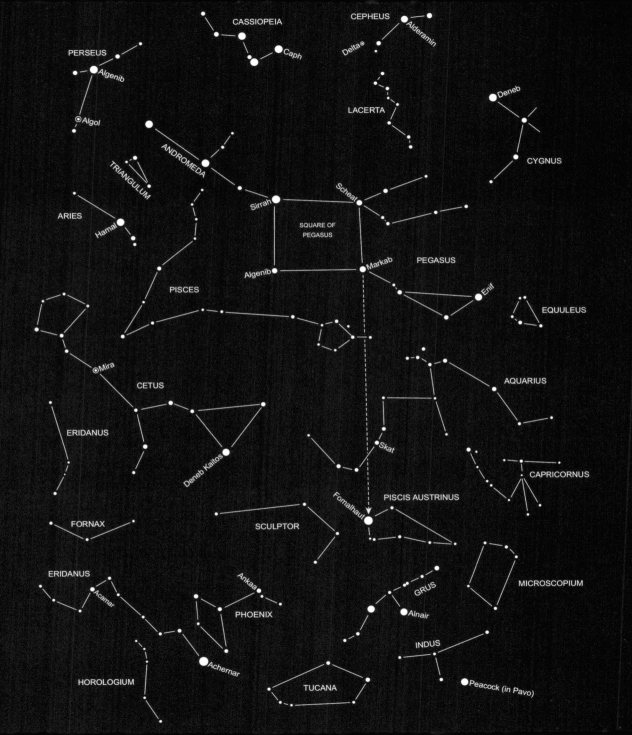

THE SOUTH CIRCUMPOLAR SKY

The area of sky depicted here shows stars which are circumpolar as seen from mid-southern latitudes and includes the constellations surrounding Octans (the Octant). Octans contains the south celestial pole, the position of which is approximately marked by the faint star Sigma Octantis (not depicted on this chart – *see Octans*). Unlike the north celestial pole, which has the comparatively bright star Polaris marking its position (*see Ursa Minor*), there is no particularly bright star marking the south celestial pole.

Although this region of sky contains many faint and obscure groups, the area is ringed by the bright stars Canopus in Carina, Achernar in Eridanus and Alpha and Beta Centauri in Centaurus, and by using these stars as guides you should have little difficulty identifying the south circumpolar constellations.

The trio of faint constellations Hydrus (the Little Water Snake), Tucana (the Toucan) and Mensa (the Table Mountain) border Octans. Along with Dorado (the Goldfish), these groups contain the two irregular galaxies the Large and Small Magellanic Clouds, the Large Magellanic Cloud situated on the border between Mensa and Dorado and the Small Magellanic Cloud located in the region between the two constellations Hydrus and Tucana.

The bright star Achernar in Eridanus (the River) is fairly easy to pick out and is located quite close to the main triangle of stars forming Hydrus. As well as being a useful guide to locating Hydrus and the other faint groups surrounding it, Achernar is also an effective finding aid for the other obscure constellations Phoenix (the

Phoenix), Horologium (the Pendulum Clock) and Reticulum (the Net). A little to the north of Phoenix we see the dim constellation Sculptor (the Sculptor) situated close to the prominent star Fomalhaut in Piscis Austrinus (the Southern Fish).

Grus (the Crane) borders Piscis Austrinus to the south with the fainter and more obscure Microscopium (the Microscope) and Indus (the Indian) located nearby. Adjacent to Indus is Pavo (the Peacock) with Telescopium (the Telescope) adjoining it. Much more prominent are the two groups Ara (the Altar) and Triangulum Australe (the Southern Triangle) both of which, along with the fainter trio Norma (the Level), Circinus (the Compasses) and Lupus (the Wolf), can be found by using the bright and distinctive pair of stars Alpha and Beta Centauri in Centaurus (the Centaur) as a location finder.

Alpha and Beta Centauri adjoin the bright constellation Crux (the Cross) which is surrounded on three sides by the southern regions of Centaurus and flanked to the south by the tiny but fairly distinctive Musca (the Fly). Situated a little

closer to the south celestial pole we find the somewhat less prominent trio Chamaeleon (the Chameleon), Apus (the Bird of Paradise) and Volans (the Flying Fish).

The small bent line of three faint stars forming Pictor (the Painter's Easel) are situated close to the bright star Canopus in Carina (the Keel) which, along with adjoining Vela (the Sail), are amongst the more prominent constellations in the area of sky covered by this chart. Equally prominent is the asterism known as the 'False Cross' which is formed from two stars in Carina and two in Vela and is often confused with the nearby constellation Crux.

BELOW: The Milky Way spans the horizon on this photograph taken by the European Southern Observatory's Very Large Telescope from the Atacama Desert in northern Chile. The dwarf irregular galaxies the Large and Small Magellanic Clouds can be seen to the left, with the glow of the magnificent globular cluster 47 Tucanae visible immediately to the upper right of the Small Magellanic Cloud.

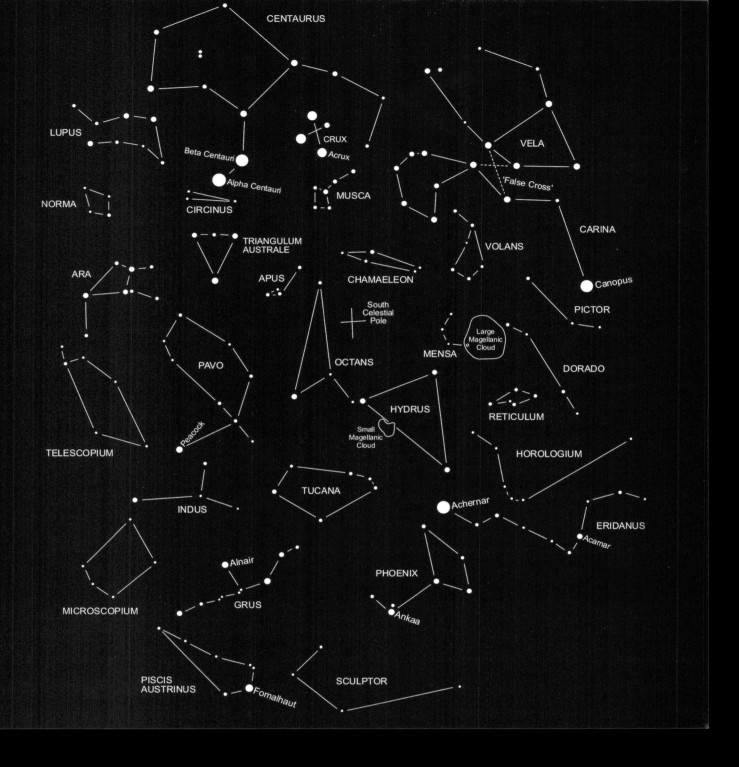

THE CONSTELLATIONS

ANDROMEDA
Andromeda

Andromeda stretches out from the northeastern corner of the Square of Pegasus and is visible in its entirety from latitudes north of 37°S. However, from locations such as South Africa, Madagascar, central South America and most of the Australian continent Andromeda will be visible low down in the northern sky and may be difficult to pick out unless the horizon is clear and free of light pollution. Observers at mid-northern latitudes, on the other hand, will see the constellation riding high in the sky, a little way to the south of the overhead point, during the autumn months.

Andromeda was the daughter of King Cepheus and Queen Cassiopeia of Ethiopia, her beauty being such that her mother Cassiopeia boasted of it far and wide, even going so far as to claim that Andromeda was more beautiful than the

THE STARS OF ANDROMEDA

Sirrah (α Andromedae) shines at magnitude 2.07 from a distance of 97 light years. The name of this star is derived from the Arabic *'surrat al-faras'* meaning 'the Horse's Navel', a reference to the time when Sirrah was considered as being a member of the adjoining constellation Pegasus (the Winged Horse).

Located a little way to the east of Sirrah is **Mirach (β Andromedae)**, a magnitude 2.07 red giant star whose light has taken 197 years to reach us. Mirach derives its name from the Arabic *'al-mi'zar'* meaning 'the Girdle' or 'the Loin Cloth'.

Orange giant **Almach (γ Andromedae)** is located at the eastern end of the main line of stars that form Andromeda. Shining from a distance of a little over 350 light years, Almach is a lovely double star with yellowish and greenish-blue components of magnitudes 2.30 and 5.10 respectively, both of which are easily resolved in small telescopes.

Delta (δ) Andromedae is a magnitude 3.27 orange giant star whose light has taken a little over 100 years to reach us.

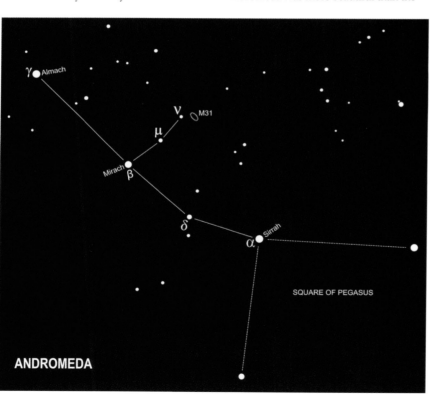

ANDROMEDA

Nereids, the nymphs of the sea. Poseidon, the god of the sea, was so angered by this that he sent a ferocious monster to terrorize the country. In an effort to resolve the situation Cepheus appealed to the gods for a solution, following which he was told to

sacrifice his daughter to the monster. Andromeda was subsequently chained to a rock to await her gruesome fate, although help was at hand in the shape of Perseus, who happened to be passing by at the time following his beheading of Medusa the Gorgon. In true Greek tradition Perseus slew the monster, released Andromeda from her chains and claimed her for his bride. The Greek goddess Athene eventually placed Andromeda in the sky, close to her husband Perseus and her parents Cepheus and Cassiopeia, where she can be seen to this day.

PAGES 30-31: The Sword Handle Double Cluster NGC 869 and NGC 884 in Perseus.

ABOVE: The Milky Way seen in the sky over an alpine village with the Andromeda Galaxy visible at left of picture.

THE GREAT ANDROMEDA GALAXY M31

Andromeda plays host to the celestial showpiece **Messier 31** (M31) or NGC 224. More popularly known as the **Great Andromeda Galaxy**, M31 can be located by following a line from Mirach through the two fainter stars **Mu (μ) and Nu (ν)** and can be seen as a faint and extended misty patch of light just to the west of Nu. This huge island universe lies at a distance of around 2.5 million light years and, with the possible exception of M33 in the constellation Triangulum, is the most distant object visible to the unaided eye.

Prior to the 20th century the Andromeda Galaxy, along with other objects of its type, had been thought of as a gas cloud located within the confines of the Milky Way. In the 1920s, however, the American astronomer Edwin Hubble proved that M31 was a galaxy in its own right and was located outside our Milky Way. By proving the existence of galaxies beyond our own, Hubble's work dramatically altered our understanding and concept of the universe. We now know that both the Andromeda Galaxy and our own Milky Way Galaxy are spiral in shape, the Andromeda Galaxy being roughly half as big again as our own. Unfortunately, however, because it lies at an angle to us, the full beauty of the Andromeda Galaxy's spiral shape is lost.

To the naked eye the Andromeda Galaxy reveals itself as a faint misty patch and has often been described as being cloud-like in appearance. As long ago as the tenth century the Persian astronomer Al-Sufi noted it as being '. . . *a little Cloud*' and in 1712 it was described by John Flamsteed, Britain's first Astronomer Royal, as a '*Nebula in Andromeda's girdle*'. When seen through binoculars M31 appears as an elliptical blur of light. Small telescopes don't show a great deal more, although larger telescopes may reveal a bright oval nucleus and some traces of the spiral arms.

It may be that the greatest satisfaction in viewing the Andromeda Galaxy is knowing what it is. It certainly impressed the English astronomer Joseph Henry Elgie who wrote that: '. . . *the great Nebula of Andromeda . . . is sunk to an appalling depth in space, and is gaseous in nature, much brighter in the center than at the edges. It is very weird-looking; a vast Solar System, probably, in process of making. Its glow seems to palpitate as I gaze at it tonight. And what an uncanny feeling one has that there is something very mysterious behind it.*'

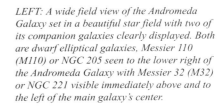

LEFT: A wide field view of the Andromeda Galaxy set in a beautiful star field with two of its companion galaxies clearly displayed. Both are dwarf elliptical galaxies, Messier 110 (M110) or NGC 205 seen to the lower right of the Andromeda Galaxy with Messier 32 (M32) or NGC 221 visible immediately above and to the left of the main galaxy's center.

OPPOSITE: A closer view of the Andromeda Galaxy with its dust lanes and spiral arms particularly prominent. The two companion galaxies M32 and M110 are seen in more detail, the elliptical shape of these two objects clearly evident.

ANTLIA
The Air Pump

The tiny and obscure Antlia is one of the 14 constellations devised by the French astronomer Nicolas Louis de Lacaille during his stay at the Cape of Good Hope in 1751/52. Created to commemorate the air pump invented by the French physicist Denis Papin, the whole of Antlia can be seen from latitudes south of 50°N.

As with many of the constellations devised by Lacaille, both Pyxis and Antlia lie in an area of sky devoid of prominent stars. The chart shows Antlia together with the neighboring Pyxis *(see below)*, the bright star Naos in the constellation Puppis being included here as a guide to tracking down these two faint groups.

THE STARS OF ANTLIA

The brightest star in Antlia is **Alpha (α) Antliae**, a magnitude 4.28 orange giant whose light has taken over 350 years to reach us.

Epsilon (ε) Antliae, is located a short way to the west of Alpha. Another orange giant star, this one shines at magnitude 4.51 from a distance of around 700 light years.

Completing the crooked line of stars forming the main part of this somewhat unimpressive constellation is magnitude 4.60 **Iota (ι) Antliae**. Located near the eastern border of Antlia, this orange giant star lies at a distance of around 190 light years.

The white giant **Theta (θ) Antliae** is the fourth brightest member of Antlia, the light from this magnitude 4.78 star having taken around 340 years to reach our planet.

A DOUBLE STAR

θ Antliae can be used as a guide to tracking down the wide double star comprising Zeta[1] (ζ[1]) and Zeta[2] (ζ[2]) Antliae. The magnitudes of these two stars are almost equally matched at 5.75 and 5.91 and, shining from a distance of over 350 light years, the pair can be found just to the west of a line from ε to θ as shown on the chart.

Binoculars will easily show both components although, if you have a small telescope, you might like to check out ζ[1]. A closer look at this star may reveal the seventh magnitude companion star which lies quite close by.

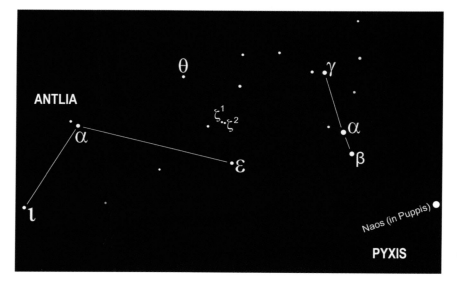

PYXIS
The Mariner's Compass

The equally small and obscure constellation Pyxis appears as nothing more than a line of faint stars to the west of Antlia and immediately to the east of the northern reaches of Puppis. Both Pyxis and Antlia can be viewed from the northern United States, central Europe, northern China and from any latitude to the south of these. Pyxis was created by Lacaille to represent the magnetic compass used by sailors and navigators, appropriately located as it is near the stern of the dismantled Argo Navis *(see the entry for Carina)*.

LEFT: Antlia plays host to a number of faint galaxies, one of which is the spiral galaxy NGC 2997 seen here. Located at a distance of around 40 million light years, NGC 2997 is face-on to us and its impressive spiral formation is well displayed.

BELOW LEFT: The Antlia Dwarf Galaxy lies at a distance of 4.3 million light years and is the faintest member of the Antlia Group of galaxies.

THE STARS OF PYXIS

Alpha (α) Pyxidis is the brightest member of this inconspicuous group, a magnitude 3.68 blue giant star shining from a distance of almost 900 light years.

Lying immediately to the south of α is **Beta (β) Pyxidis,** the magnitude 3.97 glow from this yellow giant star having taken a little over 400 years to reach us.

Rounding off the trio of stars denoting the main part of Pyxis is **Gamma (γ) Pyxidis,** a magnitude 4.02 orange giant located at a distance of just over 200 light years.

APUS
The Bird of Paradise

Another of the constellations introduced by Pieter Dirkszoon Keyser and Frederick de Houtman in the 1590s, and lying immediately to the south of Triangulum Australe, is the tiny constellation Apus, shown here with the bright pair Alpha (α) and Beta (β) Centauri and the three stars Atria and Beta (β) and Gamma (γ) Trianguli Australis as guides to locating the group.

Apus is not particularly prominent and contains only a small number of stars. It first appeared on the celestial globe produced by Petrus Plancius in 1598 following which it was depicted in the star atlas *Uranometria*, produced in 1603 by the German astronomer and celestial cartographer Johann Bayer. The constellation lies quite close to the south celestial pole, the whole of the group being visible from anywhere south of latitude 7°N.

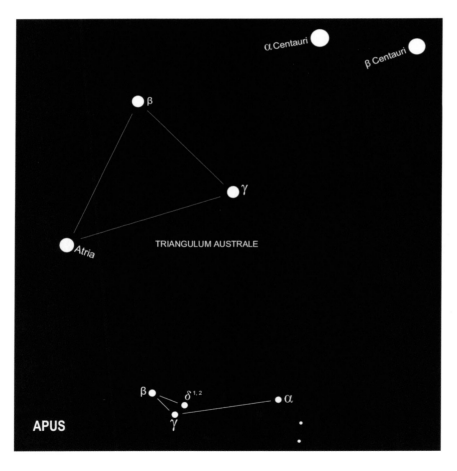

THE STARS OF APUS

The orange giant **Alpha (α) Apodis** is the brightest star in the group, shining at magnitude 3.83 from a distance of around 440 light years.

Magnitude 3.86 **Gamma (γ) Apodis** is a yellow giant star, the light from which has taken 160 years to reach us.

Beta (β) Apodis is another orange giant star shining at magnitude 4.23 from a distance of around 160 light years.

Delta (δ) Apodis is a wide optical double star, although the two stars forming it are not actually related. The brightest component is the magnitude 4.68 red giant **Delta1 (δ¹) Apodis** which lies at a distance of around 750 light years. The orange giant **Delta1 (δ²) Apodis** shines at magnitude 5.27 from a distance of around 610 light years. Both components are easily resolved in binoculars.

AQUARIUS
The Water Carrier

Immediately to the south of Pegasus we find the constellation Aquarius, the whole of which is visible from locations south of latitude 65°N. The line drawn from Scheat through Markab in the Square of Pegasus and projected southwards in order to locate the bright star Fomalhaut *(see*

Piscis Austrinus) passes immediately to the east of the 3rd magnitude star Skat in Aquarius, from where the rest of Aquarius can be identified.

Legend identifies this constellation as Ganymede, the beautiful son of King Tros of Troy. One day Ganymede was tending his father's sheep when Jupiter flew overhead disguised as an eagle. He spied Ganymede on the plains below and,

becoming infatuated with him, took the decision to kidnap Ganymede, swooping down and gathering him up in his talons. He took the boy back to Olympus, following which Ganymede became an attendant of Jupiter, one of his duties being to pour water and nectar for the gods.

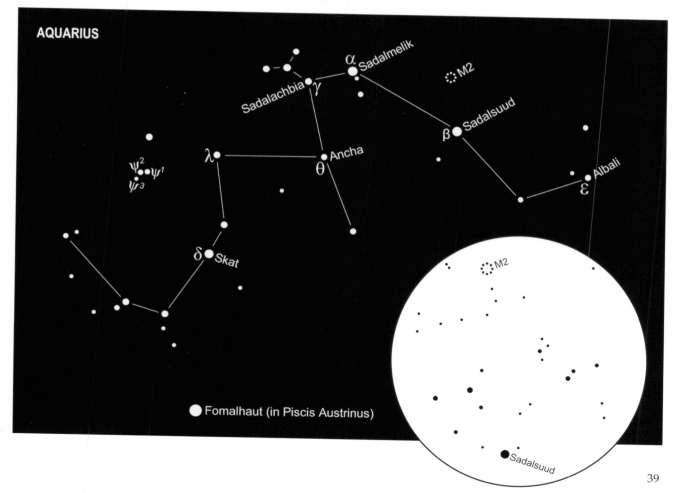

AQUARIUS

α Sadalmelik

Sadalachbia γ

Sadalsuud β

M2

Ancha θ

λ

ψ² ψ¹
ψ³

δ Skat

Albali ε

Fomalhaut (in Piscis Austrinus)

M2

Sadalsuud

THE STARS OF AQUARIUS

The brightest star in Aquarius is **Sadalsuud (β Aquarii)**, a yellow magnitude 2.90 supergiant which shines from a distance of around 540 light years. This star takes its name from the Arabic *'sa'd al-su'ud'* which may translate as 'the Luckiest of the Lucky (Stars)'.

Deriving its name from the Arabic *'sa'd al-malik'* which probably translates as 'the Lucky (Stars) of the King', magnitude 2.95 **Sadalmelik (α Aquarii)** is another yellow supergiant, the light from which has taken around 520 years to reach us.

Sadalachbia (γ Aquarii) appears to take its name from the Arabic for 'the Lucky (Stars) of the Tents', its magnitude 3.86 glow reaching us from a distance of a little over 160 light years.

Shining from a distance of 160 light years, the magnitude 3.27 star **Skat (δ Aquarii)** takes its name from the Arabic *'al-saq'* meaning 'the Shin'.

Magnitude 3.78 **Albali (ε Aquarii)** lies at the western point of Aquarius. Deriving its name from the Arabic word for 'swallower', the light from Albali set off towards us around 200 years ago.

The magnitude 4.17 orange giant **Ancha (θ Aquarii)** lies at a distance of 190 light years.

Lambda (λ) Aquarii is a red giant shining at magnitude 3.73 from a distance of 382 light years.

Located a short way to the east of Lambda, the tiny trio of stars **Psi^1 (ψ^1), Psi^2 (ψ^2)** and **Psi^3 (ψ^3) Aquarii** form a pretty naked-eye grouping although they all lie at different distances from us. Psi^1 is a magnitude 4.24 orange giant shining from a distance of 150 light years, less than half the distance of magnitude 4.41 Psi^2, the light from which set off on its journey towards us around 400 years ago. The magnitude 4.99 glow of Psi^3 reaches us from a distance of around 260 light years.

viewing it through binoculars, which will reveal it as a fuzzy ball of light with no individual stars resolved. However, even a small telescope may resolve some of the individual stars around the edge of the cluster.

GLOBULAR CLUSTER M2

Situated to the north of the star Sadalsuud, and forming a triangle with this star and the nearby Sadalmelik, is the globular cluster **Messier 2** (M2) or NGC 7089. With a diameter estimated to be around 175 light years, this is one of the largest known globular clusters, with current estimates putting it at a distance of 37,500 light years. M2 was discovered by the Italian astronomer Jean Dominique Maraldi in 1746, Messier adding it to his catalogue in 1760 where he described it as a *'Nebula without star . . .'*.

M2 has an overall magnitude of 6.3 and is easily picked up in binoculars. Using the prominent guide star Sadalsuud as your starting point you can use the finder chart, together with binoculars or a small telescope, to star hop your way to M2. Messier's description of M2 echoes the view you will get of the cluster when

OPPOSITE: One of the largest known globular clusters, Messier 2 (NGC 7089) in Aquarius is beautifully captured by the Hubble Space Telescope.

AQUILA
The Eagle

The distinctive constellation of Aquila lies on the celestial equator and as a consequence the whole constellation is visible from almost every inhabited part of the world. For northern hemisphere observers this group is one of the main summer constellations, its brightest star Altair visible fairly high up in the southern sky. Observers in the southern hemisphere get their best views of Aquila by looking high up towards the north during August and September.

Altair is the southernmost of the three bright stars that make up the Summer Triangle *(see Cygnus)*, the others being Deneb in Cygnus and Vega in Lyra. The Milky Way runs from roughly north to south, passing through Aquila on its way. As a result, the area of sky around Aquila is rich in star fields and time spent sweeping this area with binoculars on really dark, clear and moonless nights will be well rewarded.

Altair is flanked by the two slightly fainter stars Tarazed and Alshain which together form the distinctive trio that ensures Altair and its host constellation Aquila are fairly easy to recognize. The rest of Aquila can be seen extending to the southwest of these three stars, the group forming a large cross which can indeed be seen to resemble a bird in flight.

According to Greek and Roman legend Aquila depicts the eagle which carried the thunderbolts of Zeus. The constellation was of great importance to the Romans and was portrayed on many Roman coins. The Roman poet Caesius, who lived during the reign of Nero, referred to the group as the Eagle of Military Rome or the Eagle of St John.

OPEN STAR CLUSTER NGC 6709
Located at a distance of around 3,500 light years and situated to the southwest of Zeta and Epsilon near the western border of Aquila is the faint open star cluster **NGC 6709**. Shining with an overall magnitude of 6.7, NGC 6709 contains around 50 member stars, although this object may be a little difficult to pick out against the background of star fields in the Milky Way. If you carefully search the area of sky to the southwest of Zeta and Epsilon, and if the sky is dark, clear and moonless, you might just be able to pick out NGC 6709 as a faint nebulous patch of light.

OPPOSITE: Open star cluster NGC 6709 in Aquila.

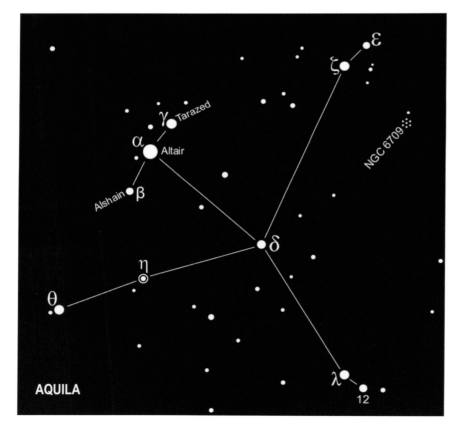

AQUILA

THE STARS OF AQUILA

The brightest star in Aquila is **Altair (α Aquilae)** which shines at magnitude 0.76 from a distance of a little under 17 light years. Altair is the brightest star in Aquila and derives its name from the Arabic '**al-nasr al-ta'ir**' meaning 'the Flying Eagle'.

Located immediately to the north of Altair is the magnitude 2.72 orange giant star **Tarazed (γ Aquilae)**, the light from which reaches us from a distance of almost 400 light years. Binoculars reveal the orange tint of Tarazed quite well.

Alshain (β Aquilae) is a yellow star of magnitude 3.71 shining from a distance of 45 light years.

Marking the central point of Aquila is **Delta (δ) Aquilae**, the magnitude 3.36 glow of which reaches us from a distance of around 50 light years.

Magnitude 2.99 **Zeta (ζ) Aquilae**, located at a distance of a little over 80 light years, is one of the two stars marking the tail of the eagle. The other is **Epsilon (ε Aquilae)**, an orange giant star shining at magnitude 4.02 from a distance of 400 light years. When viewed through binoculars, the yellow-orange glow of Epsilon contrasts nicely with the blue-white Zeta.

Located at a distance of just over 280 light years is magnitude 3.24 **Theta (θ) Aquilae.**

Eta (η) Aquilae is a Cepheid variable (*see Cepheus*) which has a range in magnitude of between 3.5 and 4.4 and a period of 7.18 days. The variability of this star was first noted by the English astronomer Edward Pigott in 1784 and, although its changes in brightness are only of less than a magnitude, you might like to try your hand at detecting them. If so, you can compare the brightness of Eta with the nearby stars Alshain (magnitude 3.71), Delta (magnitude 3.36), Theta (magnitude 3.71) and 12 Aquilae (magnitude 4.02).

Magnitude 3.43 **Lambda (λ) Aquilae** can be found near the southwestern corner of Aquila. Shining from a distance of 125 light years it is slightly closer than the orange giant 12 Aquilae, the magnitude 4.02 glow of which set off on its journey towards us around 145 years ago. These two stars are a useful guide to locating the tiny bordering constellation Scutum and are depicted on the chart of Scutum.

ARA
The Altar

Visible in its entirety from latitudes south of 22°N, the tiny constellation Ara lies immediately to the south of Scorpius, the stars Theta (θ), Eta (η) and Zeta1,2 (ζ1,2) Scorpii included here as a guide. Its location relative to Scorpius inspired the Greek poet Aratos to inform us that: *'... neath the glowing sting of that huge sign the Scorpion, near the south, the Altar hangs'.* Ara represents the altar on which Centaurus (the Centaur) is offering an animal as a sacrifice to the gods, although another account has it that Ara represents the altar created by the gods and placed among the stars by Zeus to celebrate his victory over the Titans and their leader Cronos.

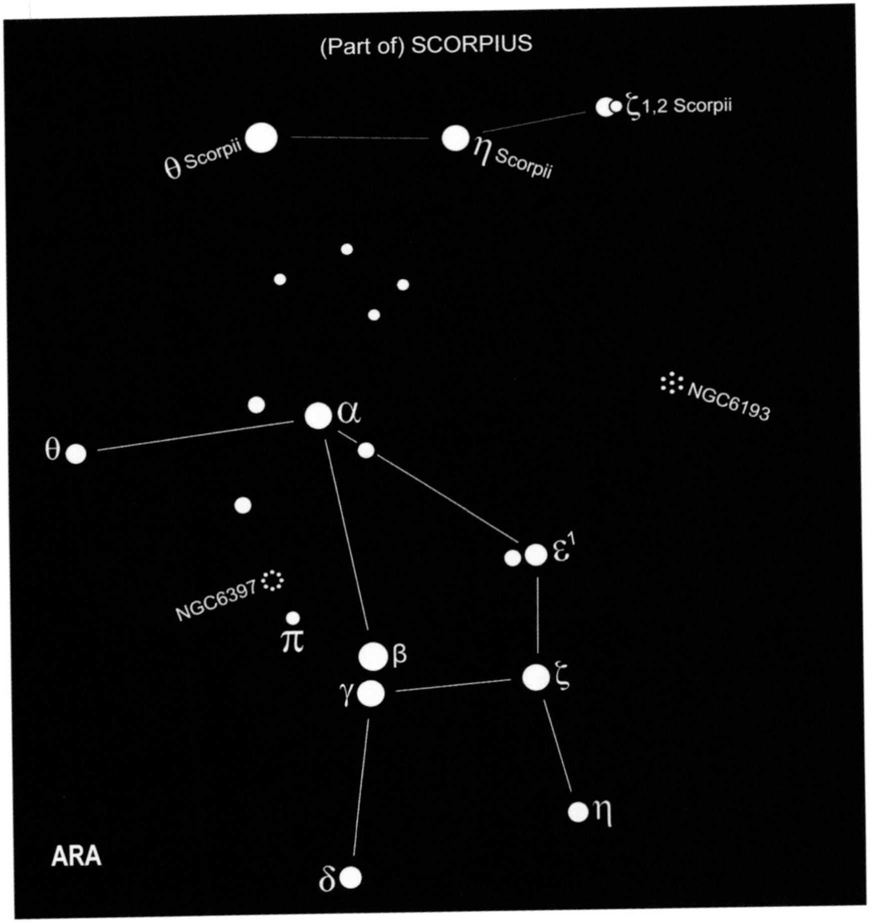

OPEN STAR CLUSTER NGC 6193
Located almost on a line from Theta (θ) Arae through Alpha, and shining with an overall magnitude of 5.2, the open star cluster NGC 6193 can be detected with the naked eye provided the sky is exceptionally dark and clear. NGC 6193 has around 30 member stars which are spread out over a wide area, making it an ideal target for the wide fields of view obtained through binoculars.

GLOBULAR CLUSTER NGC 6397
Lying roughly on a line from Beta to Theta, and located immediately to the north of Pi (π) Arae, the globular cluster NGC 6397 is another easy target for binoculars. Discovered by the French astronomer Nicolas Louis de Lacaille, it was described by him as being a *'faint star in nebulosity'*, a description which echoes what he saw through the small telescope he was using at the time. NGC 6397 shines at magnitude 5.7 and lies at a distance of around 8,400 light years, making it one of the nearest of the globular clusters.

OPPOSITE ABOVE: The open star cluster NGC 6193 is seen here enveloped by the faint nebula NGC 6188.

OPPOSITE BELOW: Globular cluster NGC 6397 imaged by the Hubble Space Telescope.

THE STARS OF ARA

Alpha (α) Arae and **Beta (β) Arae**, both of which shine at around magnitude 2.80, are the brightest stars in Ara. Alpha shines from a distance of around 260 light years, less than half the distance of the orange giant Beta, the light from which has taken a little over 600 light years to reach us.

Orange giant **Zeta (ζ) Arae** is the third-brightest star in Ara and, at magnitude 3.12, lies at a distance of 485 light years, making it somewhat more distant than magnitude 4.06 **Epsilon1 (ε1) Arae**, the light from which reaches us from a distance of 360 light years.

Located immediately to the south of Beta, albeit considerably more remote, is the magnitude 3.31 blue giant **Gamma (γ) Arae** which lies at a distance of over 1,100 light years.

The light from magnitude 3.60 **Delta (δ) Arae** set off on its journey towards us around 200 years ago, placing it somewhat closer to us than the orange giant **Eta (η) Arae** which shines at magnitude 3.77 from a distance of around 300 light years.

ARIES
The Ram

The whole of this small but well-defined constellation is visible from latitudes north of 59°S, which puts it within the visual reach of backyard astronomers across the world.

The main part of Aries takes the form of a fairly inconspicuous line of three stars which represent the head of the Ram and which lie just to the south of the neighboring constellation Triangulum, which is also depicted on the chart as a guide to location. Although Aries is not particularly outstanding in terms of visibility, it does lie in an otherwise isolated area of sky and you should have little difficulty picking it out.

Aries is associated with the legend of Jason and the Argonauts and their quest to seek out the Golden Fleece, the constellation representing the fleece which was hung on a sacred tree by King Aeetes of Colchis. The tree and its contents were guarded by a fearsome serpent, although when Jason came in search of the fleece he was helped by King Aeetes' daughter Medea to avoid the attentions of the serpent, following which he retrieved the fleece and made his escape.

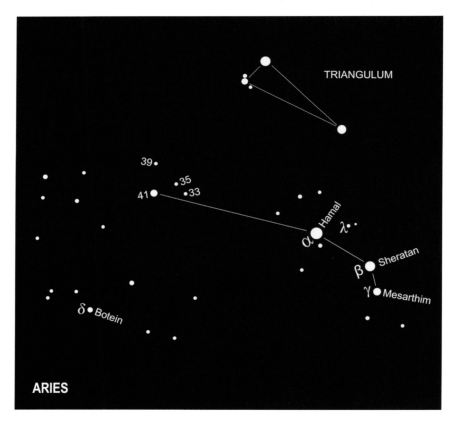

THE STARS OF ARIES

Hamal (α Arietis) is the brightest star in Aries, this magnitude 2.01 orange giant shining from a distance of around 65 light years. Hamal derives its name from the Arabic *'al-hamal'* meaning 'the Lamb'.

Sheratan (β Arietis) lies even closer, the light from this magnitude 2.64 star having set off on its journey towards us 59 years ago.

Botein (δ Arietis) derives its name from the Arabic *'al-butain'* meaning 'the Little Belly', this magnitude 4.35 yellow giant star lying at a distance of 170 light years.

Lambda (λ) Arietis lies just to the west of Hamal and is a double star with components of magnitudes 4.8 and 7.4 which can be resolved through small telescopes.

Mesarthim (γ Arietis) is a beautiful double star with magnitude 4.2 and 4.4 white components. This double is a lovely sight, even when seen through small telescopes, and is a definite target for the backyard astronomer.

MUSCA BOREALIS

Although now obsolete, and long-forgotten by most astronomers, the tiny constellation Musca Borealis (the Northern Fly) was devised by the Dutch celestial cartographer Petrus Plancius, who introduced this tiny group on a celestial globe produced by him in 1613. Formed from the four stars now designated **33**, **35**, **39** and **41 Arietis**, the original name he gave to his creation was Apes (the Bee). The constellation underwent several name changes before finally becoming Musca Borealis. During the same period there existed a southern constellation called Musca Australis (the Southern Fly), the title 'Borealis' being added to Plancius' original constellation to avoid confusion between the two. The name of Musca Australis has now been shortened to Musca (the Fly) and, although Musca Borealis no longer appears on official star charts, the stars that formed it do, and you can identify these for yourself by looking a little to the northeast of Hamal as shown here.

Detail of an 18th-century engraving of Dutch celestial cartographer Petrus Plancius instructing students in the science of navigation.

AURIGA
The Charioteer

Auriga can be seen in its entirety from latitudes north of 34°S, making it accessible to observers in South Africa, most of South America, the northern half of Australia and from all locations to the north of these. There are numerous legends attached to this prominent constellation, perhaps the best known of which is that which identifies the group as

AURIGA

THE STARS OF AURIGA

Capella (α Aurigae) is the brightest star in Auriga, and the sixth brightest star in the sky, shining at magnitude 0.08 from a distance of 43 light years. Capella derives its name from the Latin for 'she-goat' and represents the goat Amalthea who suckled the infant Zeus and who was repaid for her efforts by being placed among the stars.

A little to the southwest of Capella are her two kids, represented by **Eta (η)** and **Zeta (ζ) Aurigae**, and which are collectively known by their Latin name Haedi (the Kids). Zeta is an orange giant shining at magnitude 3.69 from a distance of 780 light years, slightly fainter and more remote that Eta, the magnitude 3.18 glow of which reaches us from a distance of a little over 240 light years. These two form a distinctive little triangle with magnitude 3.03 **Epsilon (ε) Aurigae**, a star which was known to Arabic astronomers as *al-ma'az* (the He Goat).

Deriving its name from the Arabic for 'Shoulder of the Rein-holder', **Menkalinan (β Aurigae)** shines at magnitude 1.90 from a distance of around 80 light years.

The light from magnitude 4.30 red giant **Pi (π) Aurigae** reaches us from a distance of around 730 light years.

Theta (θ) Aurigae shines at magnitude 2.65 from a distance of 165 light years.

The light from magnitude 2.69 orange giant **Hassalah (ι Aurigae)** set off on its journey towards us a little under 500 light years ago.

Myrtilus, the son of Hermes and charioteer of King Oenomaus of Elis, a district of southern Greece.

Although the main stars of Auriga seem to form a striking circlet, what appears to be the southernmost star in the group, El Nath, is actually a member of the neighboring constellation Taurus (the Bull) which borders Auriga to the south. Although El Nath depicts the tip of the bull's northern horn, this star has often been considered as belonging to either constellation. Arabic astronomers took it to be a member of Auriga, identifying it as 'the Heel of the Rein-holder'.

A TRIO OF OPEN STAR CLUSTERS M36, M37 AND M38

During the early part of the 17th century, the Italian astronomer Giovanni Batista Hodierna came across what he described as: '. . . nebulous patch(es) in Auriga', three objects which turned out to be open star

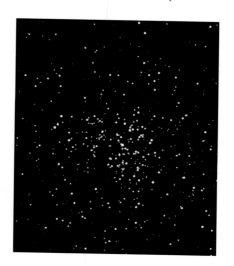

clusters and which are now known as **Messier 36** (M36) or NGC 1960, **Messier 37** (M37) or NGC 2099 and **Messier 38** (M38) or NGC 1912. Providing the sky is really dark and clear you should be able to spot these objects with binoculars, through which they will appear as faint misty patches of light. At least a small telescope will be needed in order to bring out any of their individual member stars. These clusters lie in a neat line straddling the area of sky between Theta and El Nath, the brightest of the three being M37. This object is also considered to be the finest of the trio, and was described by the German astronomer Heinrich Ludwig d'Arrest as having 'wonderful loops and curved lines of stars'. Slightly fainter is M36

and less bright still is M38. Auriga is situated against the backdrop of the Milky Way, resulting in the whole constellation being rich in stars, and carefully sweeping the area with binoculars will bring out many beautiful star fields, including those surrounding M36, M37 and M38.

LEFT: The open star cluster M37.

ABOVE: Located at a distance of around 1,500 light years, the emission/reflection nebula IC 405 surrounds the irregular variable star AE Aurigae, the energy from which star energizes the surrounding gas, causing it to glow. Also known as the Flaming Star Nebula, IC 405 shines at around 6th magnitude, and may be glimpsed in small telescopes providing the sky is really dark, clear and moonless.

BOÖTES
The Herdsman

The prominent constellation of Boötes, and its leading star Arcturus, are very noticeable in the northern spring night sky. The ruddy glow of Arcturus is fairly unmistakable, although if you do need help tracking this star down, first of all locate the familiar shape of the Plough, formed from the seven brightest stars in the constellation Ursa Major (the Great Bear) and which is located at or near the overhead point during spring evenings when viewed from mid-northern latitudes. Extending the curved line of stars in the Plough handle southwards, as shown, will take you to the brilliant orange-yellow Arcturus, from where the rest of the stars in Boötes can be traced out. Benetnash, the

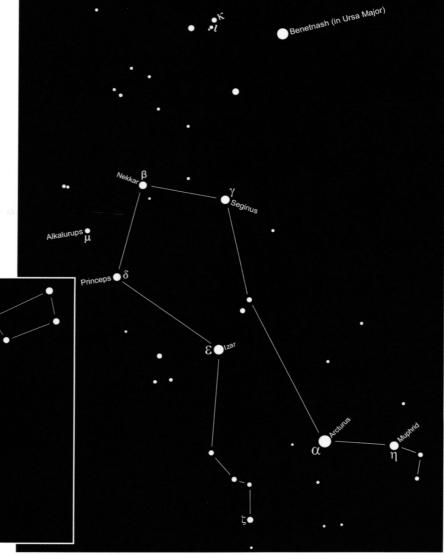

THE STARS OF BOÖTES

Arcturus (α Boötis) is the brightest star in Boötes and the 4th brightest in the entire sky, shining at magnitude –0.04 from a distance of 37 light years. It is also the most prominent member of the Diamond of Virgo, the conspicuous asterism formed from Arcturus, Spica in Virgo, Denebola in Leo and Cor Caroli in Canes Venatici and which is depicted on the chart for Coma Berenices. One of this star's main claims to fame is that it was the first to be observed in daylight, this feat having been accomplished by the French astronomer Jean-Baptiste Morin in 1635. With an obvious allusion to the nearby constellation Ursa Major (the Great Bear), Arcturus derives its name from the Greek for 'the Bear Watcher'.

Arcturus was one of the astronomer Joseph Henry Elgie's favorite stars, and he says of it: *'I have ever had the kindliest regard for golden-hued Arcturus. Glittering night after night, it is to me associated with everything that is beautiful and fair. Its changing colour, according to its altitude, is a source of never-ending delight. It is unique in its beauty and in its power of arousing kindly sentiment. To me, not only is it the harbinger of Spring, but the apotheosis of summer also.'*

Izar (ε Boötis) is a magnitude 2.35 yellow-orange giant star, its light having taken around 200 years to reach us and the color of which is fairly obvious in binoculars. Its name comes from the Arabic *'al-mi'zar'* meaning 'the Girdle', a reference to its position near the middle of Boötes.

Magnitude 3.49 **Nekkar (β Boötis)** is a yellow giant star which shines from a distance of around 225 light years.

Another yellow star is **Muphrid (η Boötis)** which derives its name from the Arabic for 'The Solitary Star of the Lancer', its magnitude 2.68 glow reaching us from a distance of 37 light years.

Seginus (γ Boötis) shines at magnitude 3.04 from a distance of 87 light years.

The light from magnitude 3.46 **Princeps (δ Boötis)** reaches us from a distance of a little over 120 light years.

Located at the southeastern corner of Boötes is the magnitude 3.78 **Zeta (ζ) Boötis,** the light from which set off towards us 175 years ago.

English accounts refer to Boötes as a Bear Driver, again due to the proximity of the constellation to Ursa Major (the Great Bear). The accounts as to the origins of this group are certainly many and varied, underlining the fact that Boötes has a somewhat mixed and uncertain origin.

A TRIO OF DOUBLE STARS IN BOÖTES

The magnitude 4.5 and 6.7 components of **Alkalurops (μ Boötis)** can be resolved through binoculars, although a small telescope will be needed to split the magnitude 4.6 and 6.6 white components of **Kappa (κ) Boötis** which can be found within the tiny triangle of stars immediately to the east of Benetnash, the end star in the handle of the Plough. Just to the south of Kappa, and located at a distance of around 95 light years, is **Iota (ι) Boötis,** its magnitude 4.8 and 6.7 components offering another target for binoculars.

'end star' in the Plough handle, is shown both on the finder chart and on the main chart of Boötes.

Boötes graces the northern skies, although portions of this constellation are visible from almost every inhabited part of the world and the whole of Boötes can be seen from South Africa and from most of Australia and South America, albeit fairly low down on the northern horizon.

Although Boötes is generally taken to represent a herdsman chasing the Great Bear (Ursa Major) around the northern sky, some accounts identify him as a hunter leading a pair of hunting dogs, depicted by the adjoining constellation Canes Venatici. Greek mythology associates the constellation with a Ploughman while early Roman stories describe the group as a Wagoner. Early

ABOVE: The brilliant Arcturus in all its glory.

CAELUM
The Graving Tool

The whole of this constellation can be seen from latitudes south of 41°N making it visible to backyard astronomers in the central United States, southern Europe and from locations further south than these. Although it contains no stars brighter than 4th magnitude, Caelum can be located with the naked eye providing the sky is fairly dark and clear

The constellation represents the burin, or graving tool, of the engraver and

is another of the constellations devised by the French astronomer Nicolas Louis de Lacaille during his stay at the Cape of Good Hope in 1751/52. Taking the form of a short zig-zag line of faint stars, it was created to fill the gap between Columba and Eridanus. As a guide to locating Caelum the three stars Beta (β) Columbae, Phakt and Epsilon (ε) Columbae in the neighboring constellation Columba are included on the chart. As is the case with many of the constellations devised by Lacaille, simply locating it can be regarded as something of an achievement.

THE STARS OF CAELUM
Alpha (α) Caeli is the brightest member of the constellation, the light from this white magnitude 4.44 star reaching us from a distance of around 65 light years.

Gamma (γ) Caeli is an orange giant star, its magnitude 4.55 glow having set off towards us around 180 years ago.

Beta (β) Caeli shines at magnitude 5.04 from a distance of 93 light years.

Delta (δ) Caeli completes the main outline of the constellation, shining at magnitude 5.07 from a distance of just over 700 light years.

CAMELOPARDALIS
The Giraffe

Although Camelopardalis is visible from all latitudes north of 5°S, the fact that it is such a dim group means that observers located around the equatorial regions will have difficulty making it out at all, unless the sky above their northern horizon is exceptionally dark, clear and free of light pollution.

One of the faintest of the constellations, the long straggling form of Camelopardalis lies in the far northern skies in an area devoid of any bright stars. So barren is this region of sky that Greek astronomers saw fit to leave it blank. It was the Dutch astronomer Petrus Plancius who invented the group in 1613 to fill the gap and, to give him credit, the group

(Part of) COLUMBA / CAELUM

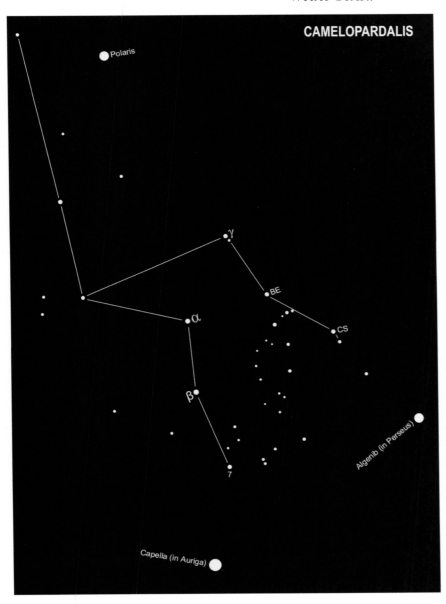

CAMELOPARDALIS

Polaris

γ

BE

α

CS

β

7

Algenib (in Perseus)

Capella (in Auriga)

THE STARS OF CAMELOPARDALIS

Beta (β) Camelopardalis is a yellow supergiant star shining from a distance of a little over 850 light years, its magnitude of 4.03 making it the brightest star in the constellation.

Alpha (α) Camelopardalis is slightly fainter, the magnitude 4.26 glow of this blue supergiant star having reached us from a distance in excess of 5,000 light years.

Gamma (γ) Camelopardalis is a white magnitude 4.59 star whose light set off towards us around 350 years ago.

does indeed resemble the character it is supposed to depict. Observers at mid-northern latitudes can find the constellation at or near the overhead point during northern winter. The head of Camelopardalis lies close to Polaris, the rest of the stars forming this obscure group extending from here southwards towards the northern reaches of the somewhat more prominent constellations Perseus and Auriga. In order to pick out the stars forming this constellation you will need a dark, clear sky which is free of moonlight. Polaris is shown here as a guide together with the stars Algenib in Perseus and Capella in Auriga.

KEMBLE'S CASCADE

Kemble's Cascade is a slightly-meandering chain of 20 or so unrelated stars which range between 5th and 9th magnitudes and which extend from a point in the sky roughly two thirds of the way from the star **CS** towards **BE**. From here it stretches for a distance roughly equal to that of five times the diameter of a full Moon in the approximate direction of the star **7 Camelopardalis**. Sweeping this area of sky with binoculars should enable you to pick out this attractive collection of stars. Also known as the Waterfall, Kemble's Cascade was named in honor of the amateur astronomer Lucian J. Kemble and is a lovely sight when viewed under dark skies.

The English astronomer Joseph Henry Elgie said of Camelopardalis that '. . . *it sprawls over a large area of sky [and] actually requires a pretty clear night before anything can be seen in it at all, so insignificant are its stars.*' Anyone who tries looking for this faint constellation will surely echo those sentiments!

The pretty chain of stars forming Kemble's Cascade meanders across the southern reaches of Camelopardalis.

CANCER
The Crab

An inconspicuous constellation comprised of stars no brighter than 4th magnitude, and which lies between the two more prominent groups Gemini and Leo, Cancer can be located by using the two stars Castor and Pollux as a guide. Located as it is a little way to the north of the celestial equator, this group can be seen in its entirety from virtually anywhere in the world.

The constellation depicts the crab commanded to crawl out of the ground by the goddess Hera to sting Hercules while he was battling the fearsome Hydra in the swamps near Lerna. Hercules responded by killing the crab, following which Hera placed the animal amongst the stars.

THE STARS OF CANCER

The orange giant **Al Tarf (β Cancri)** is the brightest star in Cancer, shining at magnitude 3.53 from a distance of around 300 light years. Its name is derived from the Arabic for 'The End', alluding to the fact that it depicts the southern foot of the crab.

Deriving its name from the Arabic for 'The Claws', **Acubens (α Cancri)** has a magnitude of 4.26 and lies at a distance of around 190 light years.

Asellus Borealis (γ Cancri) and **Asellus Australis (δ Cancri)** are the two stars occupying the central regions of the constellation, their names derived from the Latin for the 'Northern Ass' and the 'Southern Ass'. The magnitude 4.66 glow from the white supergiant Asellus Borealis set off on its journey towards us around 180 years ago, placing it somewhat further away than magnitude 3.94 Asellus Australis, a yellow giant star whose light has taken 130 years to reach us.

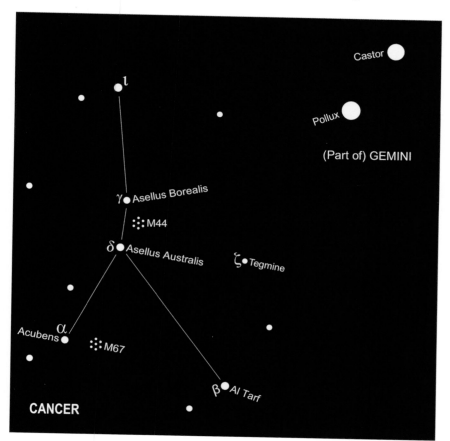

PRAESEPE

Known to astronomers by its Latin name Praesepe, but often referred to as the Beehive, Messier 44 (M44) or NGC 2632 is the brightest of the two open star clusters that can be found in Cancer. M44 shines with an overall magnitude of around 3.7 and can be glimpsed as a nebulous patch of light with the naked eye under clear, dark skies. Containing at least 350 stars, M44 is one of the nearest open star clusters to us, shining from a distance of

around 580 light years. Virtually the whole of M44 will fit into the field of view of binoculars which will bring the cluster out quite well.

The Greek astronomer Ptolemy described this object as being the '*Center of the cloud-shaped convolutions in the breast (of Cancer), called Praesepe*'. Roman mythology identifies M44 as a manger, provided as a source of food for the two nearby asses Asellus Borealis and Asellus Australis.

The astronomer Joseph Henry Elgie was sufficiently impressed with M44, and somewhat less inspired by the constellation as a whole, to say: '*A poor constellation would Cancer be if it were not for the presence of Praesepe, which now looks as cloudy to me as does Andromeda's superb nebula. Yet somehow there is a difference. But how shall I describe that difference? Shall I say that Praesepe is, to the naked eye, inanimate, whilst the Andromeda Nebula breathes and palpitates like a sentient being – the one, clay; the other, animated marble. But assist the sight with even the most modest of opera-glasses. Then does Praesepe leap into life. Its vivacity is exhilarating. Its crowded components seem to dance for the very joie de vivre, and their innumerable laughter to be infectious.*'

immediately west of the star Acubens, M67 can be tracked down fairly easily in binoculars which will reveal it as a misty patch of light. This cluster was discovered at some point during the 1770s by the German astronomer Johann Gottfried Koehler who described it at the time as *'A rather conspicuous nebula . . . '* , Charles Messier recording it in 1780 as *'A cluster of small stars . . . below the southern claw of the Crab.'*

DOUBLE STARS IN THE CRAB

Located at the northern end of the constellation we find Iota (ι) Cancri which shines with an overall magnitude of 4.03 from a distance of around 300 light years. Closer examination with a small telescope or good binoculars will reveal this to be a double star, the magnitude 4.2 and 6.6 components of which have yellowish and bluish tints.

To the west of Asellus Australis we see magnitude 4.67 Tegmine (ζ Cancri), its name derived from the Latin for 'covering' or 'shell' and which was found to be a double star by the German astronomer Johann Tobias Mayer in 1756. Located at a distance of around 82 light years, the magnitude 5.1 and 6.0 components of Tegmine are too close to each other to be resolvable in binoculars, although small telescopes will split the pair. Both components have since been found to be very close binary systems, each of which is comprised of two stars orbiting each other. However, these two individual pairs are so close together that large telescopes are required to see them.

OPPOSITE: Also known as the Beehive, M44 (NGC 2632) is one of the nearest open star clusters to us.

ABOVE: With an age estimated to be around 4 billion years, M67 (NGC 2682) is one of the oldest known open star clusters.

OPEN CLUSTER M67

One of the oldest open star clusters known to astronomers, Messier 67 (M67) or NGC 2682 contains over 500 stars and shines with an overall magnitude of 6.1 from a distance of over 2,500 light years. Located

CANES VENATICI
The Hunting Dogs

The main part of the tiny constellation Canes Venatici is visible as a pair of stars immediately to the south of the 'handle' of the Plough, the pattern of stars depicting the tail of Ursa Major (the Great Bear). Benetnash, the star at the end of the Plough handle, is shown here for reference along with the nearby bright star Arcturus in the neighboring constellation Boötes. The whole of Canes Venatici can be seen from latitudes north of 37°S although to observers in Australia, South Africa and South America the constellation will be situated fairly low down on the northern horizon.

The stars in the area of sky immediately to the south of the tail of the Great Bear were known to Arabic astronomers as 'Al Karb al Ibl', meaning 'the Camel's Burden', a description which can only allude to the stars which form Canes Venatici. The constellation we recognize today was introduced by the Polish astronomer Johannes Hevelius in the late 17th century and represents a pair of hunting dogs held by Boötes (the Herdsman), the three of them involved in an eternal pursuit of the Great Bear around the northern sky. Canes Venatici is one of only three constellations that depict dogs, the other two being Canis Major and Canis Minor.

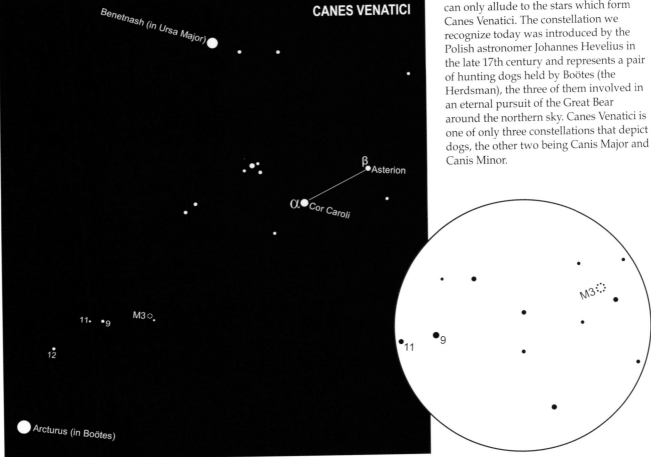

THE STARS OF CANES VENATICI

The brightest star in Canes Venatici is **Cor Caroli (α Canum Venaticorum)** or Charles' Heart which shines from a distance of around 110 light years. Along with the nearby bright stars Denebola in Leo, Spica in Virgo and Arcturus in Boötes, Cor Caroli forms the conspicuous and celebrated asterism known as the Diamond of Virgo, and which is depicted on the chart showing Coma Berenices. Cor Caroli is a double star, its magnitude 2.8 and 5.6 components being easily resolvable through a small telescope.

Exactly who Cor Caroli commemorates remains a mystery. The star is thought by some to have been named in memory of Charles I, who was executed in 1649 during the English Civil War, whilst others believe it was named after his son, Charles II, following his return to England in 1660 and his restoring of the English monarchy to the throne.

Asterion (β Canum Venaticorum) depicts the other dog from the mythological account and can be found a short distance to the northwest of Cor Caroli. Asterion shines at magnitude 4.24, its light having taken 28 years to reach us.

GLOBULAR CLUSTER M3

Canes Venatici plays host to the globular cluster Messier 3 (M3) or NGC 5272 which, like all globular clusters, lies outside our Galaxy. Located at a distance of well over 30,000 light years and glowing at magnitude 6.2, this

magnificent object was discovered by the French astronomer Charles Messier in 1764 who described it as a *'Nebula without star; center brilliant, gradually fading away; round'*. M3 was resolved into stars several years later by William Herschel who described it as *'A beautiful cluster . . .'* M3 is thought to contain around half-a-million stars compressed, as its descriptive name suggests, into a globe-shape with an actual diameter of around 180 light years.

The cluster is located almost on a direct line from Asterion, through Cor Caroli, and extended towards the bright star Arcturus in the neighboring constellation Boötes. Using binoculars, start your search from Arcturus and slowly work your way up towards Canes Venatici. Look for the faint stars 12, 11 and 9 (all in Boötes) and, once you've spotted 11 and 9, star hop your way to M3 using the detailed finder chart. Once you've managed to track it down, M3 should appear as a tiny, spherical cloud. It should be borne in mind that, when seeking out objects of this type, you need to look for a patch of luminosity rather than a point of light.

ABOVE: The superb globular cluster M3 (NGC 5272) in Canes Venatici.

BELOW: Canes Venatici is home to the spiral galaxy M63 (NGC 5055). Also known as the Sunflower Galaxy, M63 was discovered by the French astronomer Pierre Mechain in 1779.

CANIS MAJOR
The Great Dog

Located to the southeast of Orion we find Canis Major, its brightest star Sirius located by following the line of three stars forming the Belt of Orion to the southeast as shown here. Canis Major can be seen in its entirety from central Canada, northern Europe and central Russia and from any locations further south.

Legend has it that this constellation represents one of Orion's hunting dogs, the other being the adjoining Canis Minor (the Little Dog), located just to the northeast. Orion and his two dogs are pursuing a hare, depicted by the constellation Lepus and which can be found immediately to the south of Orion.

Muliphein γ

Sirius
α

Mirzam
β

ν²

12 M41

Wezen
δ

η
Aludra

Adara
ε

Furud
ζ

CANIS MAJOR

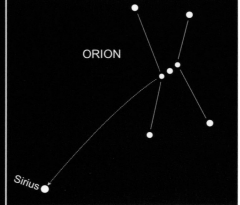

ORION

Sirius

THE STARS OF CANIS MAJOR

As well as being the brightest star in the sky, brilliant blue-white Sirius (α Canis Majoris) is also one of the closest stars to our planet, shining at magnitude −1.46 from a distance of just 8.6 light years. Sirius is so bright and prominent that the Greeks believed it to have a heating effect on the Earth, its name being derived from the Greek 'seirius' meaning 'scorching' or 'glowing'. Sirius is also known as the Dog Star, from its role as the leading star in Canis Major, the rising of this star at dawn during the time of the Greeks heralding the approach of summer and its hot (or 'dog') days.

Sirius is flanked by the two blue giant stars **Muliphein (γ Canis Majoris)** and **Mirzam (β Canis Majoris)**. Muliphein shines at magnitude 4.41 from a distance of a little over 400 light years, slightly closer than Mirzam, the magnitude 1.98 glow of which reaches us from a distance of just under 500 light years.

The light from magnitude 1.50 blue giant **Adara (ε Canis Majoris)** set off on its journey towards us around 400 years ago.

The third brightest star in Canis Major is magnitude 1.83 **Wezen (δ Canis Majoris)**, a white supergiant located at a distance of around 1,500 light years.

Aludra (η Canis Majoris) is a blue supergiant shining at magnitude 2.45 from a distance thought to be in excess of 2,000 light years.

Magnitude 3.02 **Furud (ζ Canis Majoris)** lies at a distance of around 360 light years.

The orange giant star **Nu² (ν²) Canis Majoris** shines at magnitude 3.95 from a distance of just 64 light years, and is a useful guide to locating the open star cluster M41 (see below and right).

OPEN STAR CLUSTER M41

Located a short way to the south of Sirius is the open star cluster **Messier 41** (M41) or NGC 2287. Shining with a magnitude of 4.5 this cluster is visible to the naked eye providing the sky is really dark and clear and is an easy target for binoculars. Although the discovery of M41 is credited to the Italian astronomer Giovanni Batista Hodierna during the early part of the 17th century, it may have been observed by the Greek philosopher Aristotle as long ago as the fourth century BC. When Charles

Messier added it to his catalogue in 1765 he described it as: '*A cluster of stars below Sirius; this cluster appears nebulous in an ordinary telescope . . . it is nothing more than a cluster of small stars*'. Located at a distance of 2,400 light years, M41 contains about 100 stars and measures around 25 light years across.

To locate M41, use binoculars to check out the area of sky a little way to the south of Sirius. The M41 cluster forms a triangle with Sirius and the nearby star Nu², as shown on the chart. Another guide

to location is the faint star 12 Canis Majoris on the southern edge of the cluster. When searching for M41, remember to look for a misty patch of light rather than a star-like point.

ABOVE: Open star cluster M41 (NGC 2287).

PAGE 62: Another of Canis Major's jewels is NGC 2359. Also known as Thor's Helmet, this huge emission nebula has a diameter of around 30 light years and lies at a distance in excess of 12,000 light years.

CANIS MINOR
The Little Dog

The tiny constellation of Canis Minor lies close to the celestial equator and can be seen from every inhabited part of the world. Located a little to the northeast of Canis Major, it can be readily identified by its leading star Procyon.

Although generally recognized as being one of Orion's hunting dogs, the group is also identified with the gigantic Teumessian Fox, a creature that, according to legend, was destined never to be caught. This huge fox was one of the children of Echidna, a fearsome being who was half-woman, half-snake and known as the 'Mother of All Monsters'. The Theban general Amphitryon was given the seemingly-impossible task of destroying the Teumessian Fox to which end he fetched the magical dog Laelaps who had the distinction of catching every creature he chased. However, the abilities of these two creatures were mutually excluding and Zeus, seeing the situation as being irresolvable, turned them both into stars and placed them in the sky where they can be seen to this day. Laelaps is identified with the constellation Canis Major.

THE STARS OF CANIS MINOR

Procyon (α Canis Minoris) and Sirius (in the nearby constellation of Canis Major) are two of the closest stars to our solar system. Procyon shines from a distance of 11.4 light years, a couple of light years or so further away than Sirius. Procyon derives its name from the Greek for 'the One Preceding the Dog', alluding to the fact that Canis Minor rises before the neighboring celestial dog Canis Major. To the naked eye Procyon looks distinctly white, although a good pair of binoculars may reveal its slightly yellowish tint.

Located at a distance of around 170 light years **Gomeisa (β Canis Minoris)** is a white star with a magnitude of 2.89 and a true luminosity of some 250 times that of our Sun. Gomeisa derives its name from the Arabic **'al-ghumaisa'** meaning 'the Dim, Watery-eyed or Weeping One', a title that was originally applied by Arabic astronomers to the constellation as a whole and one which illustrates the point that star names are quite often nothing if not imaginative!

Gamma (γ) Canis Minoris shines from a distance of over 300 light years and has a distinctly orange-yellow tint which can be spotted through binoculars.

The constellation of Canis Minor is quite small and contains little of interest to the backyard astronomer, although the area of sky containing Gomeisa, Gamma, Epsilon (ε) and Eta (η) is quite pretty and well worth a look through binoculars.

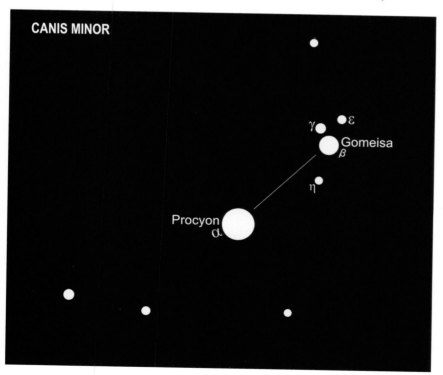

CANIS MINOR

γ ε

Gomeisa
β

η

Procyon
α

CAPRICORNUS
The Goat

Located to the southeast of the constellation Aquila and its leading star Altair (as shown on the chart of the Northern Summer / Southern Winter Sky on page 25) is the constellation Capricornus. Visible in its entirety from anywhere south of latitude 62°N, Capricornus is always low down in the southern sky as seen by those at mid-northern latitudes, although observers in the southern hemisphere will find the group riding high in the northwestern sky during October evenings.

Capricornus is one of the 48 constellations listed by the Greek astronomer Ptolemy during the second

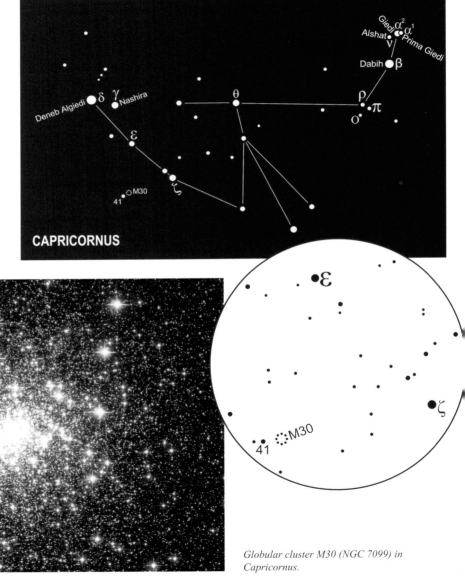

Globular cluster M30 (NGC 7099) in Capricornus.

THE STARS OF CAPRICORNUS

Deriving its name from the Arabic **'dhanab al-jady'** meaning 'the Kid's Tail', **Deneb Algiedi (δ Capricorni)** is the brightest star in Capricornus. Deneb Algiedi is found at the eastern end of the constellation, shining at magnitude 2.85 from a distance of 39 light years.

Nashira (γ Capricorni lies just to the west of Deneb Algiedi, the magnitude 3.69 glow of this white giant star having taken just over 150 years to reach us.

To the southwest of Deneb Algiedi we find the two stars **Epsilon (ε)** and **Zeta (ζ) Capricorni** which can be used as a guide to locating the globular cluster M30 (see below and left). Epsilon shines at magnitude 4.51 from a distance of over 1,000 light years, somewhat more remote than Zeta, a magnitude 3.77 yellow supergiant whose light has taken around 380 years to reach us.

The yellow giant **Giedi (α² Capricorni)**, shining with a magnitude of 3.58 from a distance of 106 light years, takes its name from the Arabic **'al-jady'** meaning 'the Kid'. Giedi forms a wide double with the nearby **Prima Giedi (α¹ Capricorni)**, a yellow magnitude 4.30 supergiant located immediately to its northwest. Giedi and Prima Giedi can be resolved with the naked eye and binoculars bring them out well. However, these two stars are not actually related, their proximity to each other being nothing more than a line of sight effect. The light from Prima Giedi reaches us from a distance of around 560 light years, putting it over five times as far away as Giedi.

The light from magnitude 4.77 **Alshat (ν Capricorni)** set off towards us around 250 years ago.

The orange giant **Dabih (β Capricorni)** is a magnitude 3.05 star shining from a distance of 340 light years and located short way to the south of Alshat and Giedi. Dabih is a binary star with components of magnitudes 3.05 and 6.10 and which offers an easy target for binoculars.

Theta (θ) Capricorni) shines at magnitude 4.08 from a distance of a little over 160 light years.

Forming a tiny triangle with **Rho (ρ)** and **Pi (π) Capricorni**, a little to the southeast of Dabih, is the double star **Omicron (o) Capricorni**, the magnitude 5.94 and 6.74 components of which can be resolved in a small telescope.

cluster shines with an overall magnitude of 7.2 from a distance in excess of 25,000 light years and was discovered by Charles Messier in 1764, being described by him as: *'Nebula discovered near 41 Capricorni. Round, contains no star. Found from (Zeta) Capricorni.'*

Subsequent observers, who were better equipped than Messier, were able to resolve more detail in M30, including William Henry Smyth who described this object as *'A fine, pale white cluster'* and Thomas William Webb who recorded M30 as being: *'Moderately bright, beautifully contrasted . . . comet-like . . . with higher powers, resolvable.'*

Tracking M30 down can be a little difficult, although two of the guide stars used by Messier are depicted on the finder chart which should help in your search. To find the cluster, first of all locate the two stars Epsilon and Zeta, following which you can star hop your way to M30 through the field of faint stars shown on the chart. M30 lies immediately to the west of the faint star 41 Capricorni and will appear as little more than a diffuse ball of light in binoculars or small telescopes. However, slightly larger telescopes, or very powerful binoculars, may enable you to resolve some of its outlying member stars.

character notable for having the hindquarters, legs and horns of a goat and the tail of a fish. Pan assisted Zeus in his battle with the monster Typhon, for which Zeus showed his gratitude by placing Pan in the sky where we see him today in the form of Capricornus.

GLOBULAR CLUSTER M30

The globular cluster **Messier 30** (M30) or NGC 7099 lies near the southeastern corner of Capricornus, situated immediately to the west of the faint star 41 Capricorni and forming a triangle with nearby Zeta and Epsilon Capricorni. This

CARINA
The Keel

Lying well to the south of the celestial equator is the constellation Carina, the leading star of which can be located by using some of the stars in Orion as a guide. Extending the line from Mintaka through Saiph southwards, as shown, will eventually bring you to brilliant Canopus. This is the brightest star in Carina and lies at the western end of the constellation, the rest of Carina trailing off towards the east.

Parts of Carina can be seen from the southern United States, southern Europe and southern China although it is only from latitudes south of 15°N that the whole of this constellation can be observed.

The list of 48 constellations drawn up by the Greek astronomer Ptolemy during the second century included Argo Navis (the Ship Argo), a large and sprawling group representing the ship in which Jason and the Argonauts journeyed to Colchis in their quest for the Golden

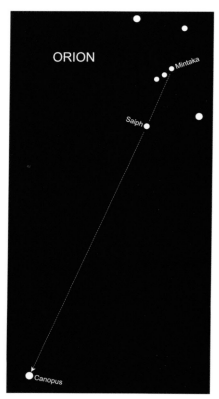

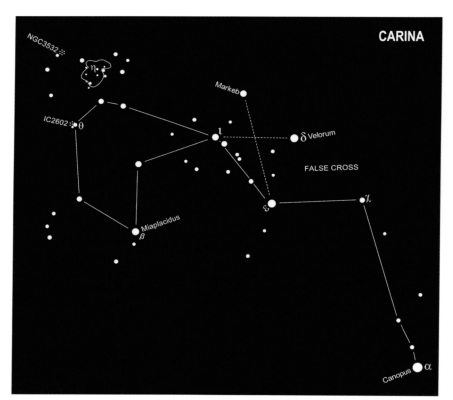

Fleece. The sheer size of Argo Navis prompted the French astronomer Nicolas Louis de Lacaille to divide it into the separate constellations Carina, Puppis (the Poop or Stern) and Vela (the Sail), all three of which have survived and made their way onto modern star charts.

OPEN STAR CLUSTER IC 2602
Located at a distance of nearly 500 light years and containing around 60 member stars, the open star cluster **IC 2602** is an excellent target for the backyard

THE STARS OF CARINA

Canopus (α Carinae) is a white supergiant star shining at magnitude −0.72 from a distance of over 300 light years. With a true luminosity of over 10,000 times that of our own Sun, Canopus is the second brightest star in the entire sky.

The light from magnitude 1.67 **Miaplacidus (β Carinae)** reaches us from a distance of around 110 light years. Although the origins of this name are uncertain, it may derive from **'Mi'ah'**, the plural form of the Arabic for water.

The magnitude 1.86 orange giant **Epsilon (ε) Carinae** lies at a distance of over 600 light years.

Chi (χ) Carinae shines at magnitude 3.46 from a distance of around 450 light years.

Theta (θ) Carinae, located near the eastern border of the constellation, is a magnitude 2.74 blue-white star and is the brightest member of the open star cluster IC 2602 (see below).

astronomer. With an overall magnitude of 1.9 this cluster is easily visible to the unaided eye. IC 2602 is also known as the Theta Carinae Cluster due to the fact that its brightest member is the star Theta Carinae. The cluster was discovered by Nicolas Louis de Lacaille in 1751 during his stay in South Africa cataloguing the southern stars, and at the time he was dividing up the constellation Argo Navis (see above). Because IC 2602 occupies an

area of sky larger than the full Moon, the wide field of view obtained with binoculars make them the ideal instrument with which to view this lovely cluster.

OPEN STAR CLUSTER NGC 3532

Lying a little way from the Eta Carinae Nebula is the open star cluster **NGC 3532**. Containing in the region of 150 member stars, this magnificent object shines at 3rd

BELOW: The bright open cluster NGC 3532 in Carina. Also known as the Wishing Well Cluster and the Football Cluster, this object is an easy naked-eye target.

magnitude from a distance of around 1,400 light years. NGC 3532 occupies an area of sky greater than the full Moon, the wide fields of view obtained with binoculars being a definite asset when viewing this wonderful object.

THE ETA CARINAE NEBULA

Another of the discoveries made by Nicolas Louis de Lacaille, the Eta Carinae Nebula is thought to lie at a distance of around 8,000 light years and takes the form of a vast cloud of gas surrounding the variable star **Eta (η) Carinae**. Also known as the Carina Nebula, this huge and highly luminous object rivals the Orion Nebula for sheer splendor and is well worth observing, even with binoculars or a small telescope.

When first catalogued, the star Eta Carinae was of 3rd magnitude, although it was seen to vary considerably over the following years, reaching magnitude –0.8 in 1843, before sinking to 6th magnitude obscurity by 1868 where it remains today. Material ejected from the star, or from dust clouds surrounding it, are thought to play a role in the variability of Eta Carinae.

THE FALSE CROSS

Magnitude 2.21 **Iota (ι) Carinae** is a white supergiant star, the light from which set off towards us over 750 years ago. A little way to the west of Iota is the orange giant **Epsilon (ε) Carinae**, the magnitude 1.86 glow of which reaches us from a distance of 600 light years. These two stars, along with Delta (δ) and Markab in the adjoining constellation Vela, make up the False Cross, an asterism which is often confused with the constellation Crux (the Cross or Southern Cross) which lies some way to the east of Carina. However, there are significant differences between the two, Crux being smaller and generally more distinct than the False Cross.

The Eta Carinae Nebula.

CASSIOPEIA
Cassiopeia

The prominent 'W' or 'M' formation of stars known as Cassiopeia graces the far northern skies and represents the mythological Queen Cassiopeia, wife of King Cepheus and mother to Andromeda, the beautiful maiden rescued by Perseus. The entire constellation can be seen from locations to the north of latitude 12°S, with portions of Cassiopeia visible from Australia, South Africa and all but the southernmost reaches of South America. Observers at mid-northern latitudes should have no difficulty locating this constellation which stands out well at or around the overhead point during autumn evenings.

For observers in Great Britain and northern Europe, Cassiopeia is a circumpolar constellation, which means that it never sets as seen from these latitudes. To find Cassiopeia, first of all identify the two end stars in the 'bowl' of the Plough. Also known as 'the Pointers', these are the two stars that are generally used to locate Polaris, the Pole Star, in Ursa Minor. Following the imaginary line from Merak, through Dubhe and past Polaris as shown overleaf, will eventually lead you to the star Caph in Cassiopeia.

The constellation lies within the Milky Way and on a really clear night you should be able to see around 50 naked-eye stars within the group, and binoculars will reveal many more. Their combined light produces the effect we call the Milky Way (*see Glossary*), the full visual effect of which, although generally lost to city-dwellers, can be a superb sight when viewed under a really dark and moonless sky.

OPEN STAR CLUSTER M52

Discovered by Charles Messier in 1774, the open star cluster **Messier 52** (M52) or NGC 7654 is just one of several clusters found in Cassiopeia. Shining with an overall magnitude of 7.3, M52 contains around 200 stars and lies at a distance in excess of 5,000 light years. Visible in binoculars or a small telescope, M52 can be found close to the star 4 Cassiopeiae, lying more or less on a line taken from Shedar through Caph. You can track the

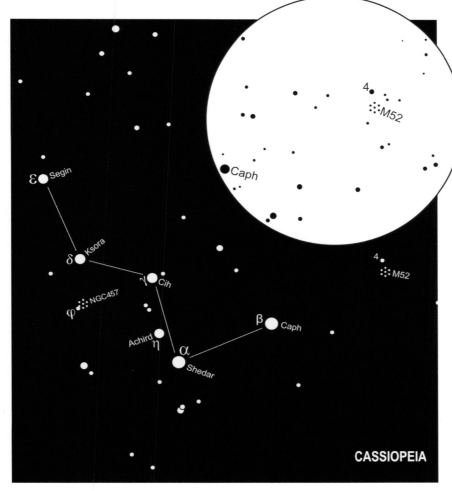

CASSIOPEIA

THE STARS OF CASSIOPEIA

The brightest star in Cassiopeia is **Cih (γ Cassiopeiae)**, the light from which has taken around 550 years to reach us. Although Cih generally shines at magnitude 2.15, this star is slightly variable and prone to sudden and unpredictable increases in brightness.

Shedar (α Cassiopeiae) derives its name from the Arabic *'al-sadr'*, meaning 'the Breast'. This is an orange giant star shining at magnitude 2.24 from a distance of around 220 light years. Binoculars will show the orange tint of Shedar quite well.

Caph (β Cassiopeiae) is a white giant, its magnitude 2.28 glow having taken just 55 years to reach us.

Even closer to us is **Achird (η Cassiopeiae)** which shines at magnitude 3.46 from a distance of just 19.4 light years.

The knee of Cassiopeia is depicted by **Ksora (δ Cassiopeiae)**, a magnitude 2.66 star located at a distance of around 100 light years. An alternative name for this star is Ruchbah, both names being derived from the Arabic *'rukbat dhat al-kursiy'* meaning 'the Knee of the Lady in the Chair'.

Magnitude 3.35 **Segin (ε Cassiopeiae)** completes the distinctive shape of Cassiopeia, its light having set off towards us around 425 years ago.

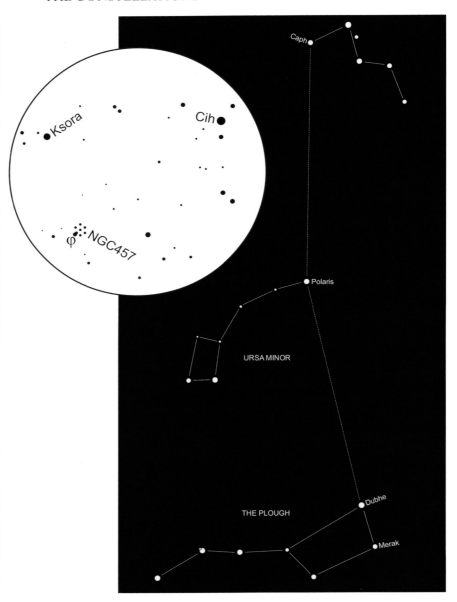

cluster down by star hopping through the field of stars from Caph as shown on the finder chart. During your search, remember to look for a small patch of light rather than an actual gathering of stars. Once the cluster is located, binoculars will reveal it as a misty cloud, although even a small telescope will bring out a number of individual stars within this fine cluster.

NGC 457 – THE OWL CLUSTER
The open star cluster **NGC 457** was discovered by William Herschel in 1787 and shines with an overall magnitude of

6.4 from a distance of between 7,500 and 10,000 light years. Two relatively bright stars, one being the magnitude 4.95 Phi (φ) Cassiopeiae, appear to lie on the

edge of the cluster. These two stars stand out quite well and give the distinct impression of being a pair of eyes staring back at the observer, which has led to NGC 457 being nicknamed 'The Owl Cluster'.

NGC 457 is located adjacent to the star Phi and forms a triangle with the two brighter stars Ksora and Cih. These three stars are depicted on the finder chart, which will enable you to track NGC 457 down by working your way through the field of fainter stars in the immediate area. As is often the case with objects of this type, binoculars will show the cluster as a distinct patch of light, although a small telescope should reveal a couple of dozen or so of its individual member stars.

ABOVE: Open star cluster M52 (NGC 7654).

ABOVE RIGHT: The Owl Cluster (NGC 457).

RIGHT: A shooting star passes the constellation Cassiopeia.

PAGE 72: Two emission nebulae, the Heart Nebula (IC1805 - right) and Soul Nebula (IC1848 - left) in Cassiopeia.

CENTAURUS
The Centaur

Centaurus depicts the legendary Chiron, the half-man and half-horse son of Cronos, king of the Titans, and the sea nymph Philyra. This is the ninth largest constellation in the sky and contains over 100 naked-eye stars. Centaurus is visible in its entirety from anywhere south of latitude 25°N, although from mid-European latitudes the constellation is largely hidden from view.

R CENTAURI – A MIRA-TYPE VARIABLE STAR
The reddish tint of the long-period Mira-type *(see Cetus)* variable star **R Centauri** is easily identifiable in binoculars. Shining from a distance in excess of 1,000 light years and located a little way to the north of the line from Alpha to Beta Centauri, R Centauri varies between 5th and 11th magnitude over a period of around 550 days. R Centauri may well be out of view when you first look for it, although if you keep watching it will eventually reappear on its approach to maximum brightness, from where you will be able to monitor the star as it increases and then decreases in magnitude. Most of its cycle of variability can be followed through binoculars, although you will need a small telescope (or very large binoculars) to keep it in view when at or near its dimmest. This is a star well worth the attentions of the backyard astronomer!

THE PEARL CLUSTER - NGC 3766
The Milky Way passes through the southern reaches of Centaurus and time spent sweeping the area with binoculars will be well rewarded. As well as the numerous rich and attractive star fields you will come across, you may also pick out a number of open star clusters, an excellent example being **NGC 3766**. Also known as the Pearl Cluster, NGC 3766 is located immediately to the north of the star Lambda (λ) Centauri and lies within the same binocular field of view. Containing around 40 member stars, NGC 3766 was discovered by the French astronomer Nicolas Louis de Lacaille during his time in South Africa cataloguing the southern stars. NGC 3766 shines at magnitude 5.3 and can be made out with the naked eye providing the sky is really dark and clear. Most of the stars

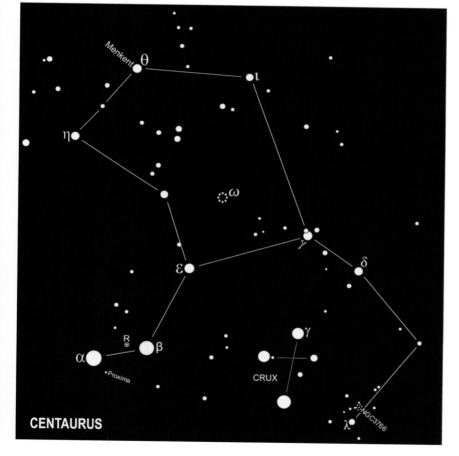

CENTAURUS

THE STARS OF CENTAURUS

Alpha (α) Centauri is the brightest star in Centaurus and, shining from a distance of 4.3 light years, the closest of the naked-eye stars. An alternative name for this star is Rigil Kentaurus, derived from the Arabic *'rijl qanturis'* meaning 'the Centaur's Foot'. Alpha Centauri is a binary star with yellow magnitude 0 and 1.2 components which orbit each other over a period of 80 years.

Both components of Alpha Centauri are easily resolvable in small telescopes, unlike the much fainter **Proxima Centauri**, the location of which is shown on the chart. This dim 11th magnitude red dwarf star is probably gravitationally linked to the Alpha Centauri system, orbiting the main pair of stars over a period of many thousands of years. The name of this star is taken from the Latin *'proxima'* meaning 'nearest' or 'closest', reflecting the fact that, at a distance of just 4.24 light years, Proxima Centauri is the nearest star to our solar system.

Beta (β) Centauri is a blue giant star situated a little to the west of Alpha Centauri. The magnitude 0.60 glow of Beta reaches us from a distance of around 350 light years. Alternative names for this star include Agena, a title bestowed on it by the American astronomy writer and cartographer Elijah Hinsdale Burritt. The meaning of the word is unclear although it may be derived from the Latin *'genu'* meaning 'knee (of the Centaur)'.

Alpha and Beta Centauri are known collectively as the Southern Pointers. This alludes to the fact that a line drawn from

Alpha through Beta leads to a point very close to **Gamma (γ) Crucis**, the star at the northern point of the adjoining constellation Crux, the distinctive cruciform shape of which is surrounded on three sides by the southern regions of Centaurus.

Centaurus is home to many bright stars including the orange giant **Menkent (θ Centauri)** which shines at magnitude 2.06 from a distance of 59 light years. Located at the northern end of the Centaur, its name is probably derived from the Arabic word *'mankib'* meaning 'the Shoulder (of the Centaur)'.

The two stars **Iota (ι) and Eta (η) Centauri** lie to either side of Menkent. Magnitude 2.75 Iota is located at around the same distance as Menkent, while somewhat further away is Eta, the light from this magnitude 2.33 star having taken a little over 300 years to reach us.

One of the southernmost members of Centaurus is magnitude 3.10 **Lambda (λ) Centauri**, the light from which set off towards us over 400 years ago.

Magnitude 2.20 **Gamma (γ) Centauri** lies at a distance of 130 light years.

Delta (δ) Centauri shines at magnitude 2.58 from a distance of a little over 400 light years.

Epsilon (ε) Centauri is a blue giant star, the magnitude 2.29 glow of which has taken 425 years to reach us.

in the cluster are bluish in colour, although there are a couple of red giant stars which you may be able to pick out.

OMEGA CENTAURI

Omega (ω) Centauri (NGC 5139) is one of the finest examples of all the globular clusters and is a prominent naked-eye object. Globular clusters are vast spherical collections of stars located in the area of space surrounding the Galaxy *(see Glossary)*, unlike open clusters, which have no really well-defined shape and are found within the main galactic plane. In contrast to the young hot stars from which many open clusters are formed, globular clusters are made up of old stars with little or none of the nebulosity sometimes seen in open clusters.

Shining from a distance of nearly 16,000 light years, and with a diameter in excess of 150 light years, Omega Centauri is the largest globular cluster in our Milky Way Galaxy and contains several million individual stars. Omega Centauri can be tracked down by extending a line from

Beta Centauri, through Epsilon and on nearly as far again, binoculars revealing the cluster as a large nebulous patch with even a small telescope bringing out many individual member stars.

OPPOSITE: The Pearl Cluster NGC 3766.

ABOVE: Omega Centauri (NGC 5139).

When viewed without optical aid this object takes on the appearance of a nebulous star, and was given the name Omega Centauri by the German astronomer and celestial cartographer Johann Bayer who included it in his star atlas *Uranometria*, published in 1603 *(see Glossary* – Star Names). Because this object appeared to be the 24th brightest star in

Centaurus, the 24th letter of the Greek alphabet was used to identify it. Although Bayer saw it as nothing more than a hazy 4th-magnitude star, the advent of the telescope eventually revealed its true nature. However, in spite of Omega Centauri not actually being a star, the name bestowed upon it by Bayer has remained in use.

CEPHEUS
Cepheus

Depicting King Cepheus, the husband of Queen Cassiopeia and father of the beautiful Princess Andromeda, is the large but relatively inconspicuous constellation Cepheus. Following an imaginary line from Shedar through Caph, both in the neighboring constellation Cassiopeia, will lead you to Cepheus which, although possessing a fairly distinctive shape, can not be said to particularly resemble the character it is supposed to represent.

For observers at mid-northern latitudes Cepheus is a circumpolar constellation and can be seen at or near the overhead point during September and

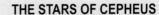

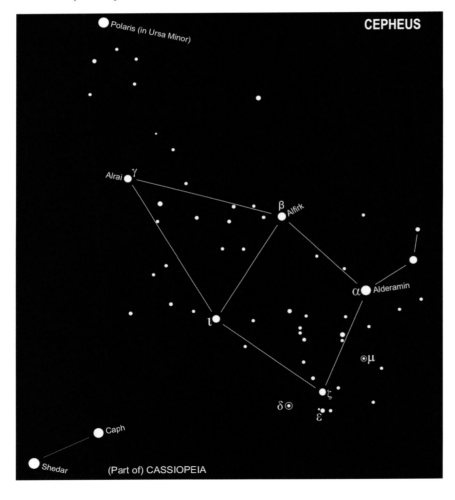

THE STARS OF CEPHEUS

Alderamin (α Cephei) marks the right shoulder of Cepheus and is the brightest star in the constellation, shining at magnitude 2.45 from a distance of around 50 light years.

The waist of Cepheus is marked by **Alphirk (β Cephei)**. Located at a distance of nearly 700 light years, Alphirk is a double star, its bluish and white magnitude 3.5 and 8.0 components resolvable through a small telescope.

Marking the left knee of Cepheus, the light from magnitude 3.21 orange giant **Alrai (γ Cephei)** set off on its journey towards us around 45 years ago. The yellow-orange tint of Alrai can be detected with a pair of binoculars.

Magnitude 3.50 **Iota (ι) Cephei** is another orange giant star, the light from which reaches us from a distance of 115 light years.

The orange supergiant **Zeta (ζ) Cephei** completes the main outline of Cepheus, the magnitude 3.39 glow of this star reaching us from a distance of over 800 light years.

Epsilon (ε) Cephei shines at magnitude 4.18 from a distance of 84 light years.

October evenings. At this time Cepheus is visible in its entirety from all latitudes north of the equator, although stargazers on or very close to the equator will see the group very low over the northern horizon. To observers further south most if not all of Cepheus is permanently hidden from view.

THE GARNET STAR

Near the southern border of Cepheus we find red supergiant star **Mu (μ) Cephei.** Located at a distance of between 2,500 and 3,000 light years, this is one of the most luminous stars known. It is also one of the largest stars visible to the naked eye and is so huge that, if it was put in the place of our own Sun, its surface would extend out to well beyond the orbit of Jupiter. The color of Mu is very distinctive, William Herschel describing it as being *'a very fine deep garnet color'*, leading to the star being popularly known as Herschel's 'Garnet Star'.

Mu Cephei is a semi-regular variable, its magnitude ranging between around 3.40 and 5.10 over a period of between 2 and 2½ years. You can compare its

brightness with the two nearby stars Zeta (magnitude 3.39) and Epsilon (magnitude 4.18). Those observers at mid-northern latitudes who want to observe the variations in brightness of Mu Cephei are fortunate in that this star is circumpolar. Consequently, it can be seen all year round and its cycle of variability continuously monitored. However, backyard astronomers in or around the equatorial regions may have to content themselves with simply checking out the conspicuous colour of Herschel's Garnet Star.

LEFT: Herschel's Garnet Star.

BELOW: Discovered by William Herschel in 1798 and located near the border between Cepheus and the neighboring constellation Cygnus is the Fireworks Galaxy (NGC 6946).

DELTA CEPHEI

Delta (δ) Cephei is one of the most famous and best-known variable stars in the entire sky and the star after which a whole class of variable star is named. The variations of Delta Cephei were discovered in 1784 by the astronomer John Goodricke, since which time hundreds of Cepheid variables have been found. Cepheids are short-period pulsating variables that are large and very luminous and can be seen over immense distances, their periods ranging between as little as a day to around a couple of months or more. Delta Cephei itself varies between magnitudes 3.5 and 4.3 over a period of 5.367 days and, as with nearby Mu Cephei, its brightness can be compared with the two nearby stars Zeta and Epsilon.

In 1912, after examining photographs of Cepheids in the Small Magellanic Cloud, the American astronomer Henrietta Swan Leavitt announced an important relationship between the true luminosities and periods of Cepheid variables. She noticed that Cepheids with shorter periods were always fainter than those whose periods of variability were longer. Her conclusion was that, because all the Cepheids under examination were at more or less the same distance from us, their apparent brightness seemed to reflect their true brightness and that those with longer periods had greater actual luminosities. In other words, the longer the period of a Cepheid variable, the brighter the star.

The upshot was that, once the true brightness of a Cepheid had been worked out, this could be compared to how bright the star actually appeared in the sky and the distance to the Cepheid calculated. This relationship between true and apparent brightness allowed astronomers to calculate the distances to Cepheids and therefore the distances to external galaxies in which Cepheids were observed. Following the observation of Cepheids in nearby galaxies by the American astronomer Edwin Hubble in the 1920s, he was able to prove that these huge stellar gatherings were systems well outside our own.

BELOW: The Cave Nebula, a mixture of emission, reflection and dark nebulosity in Cepheus.

OPPOSITE: The Elephant's Trunk Nebula is a region of gas and dust contained within the much larger emission nebula IC1396 in Cepheus.

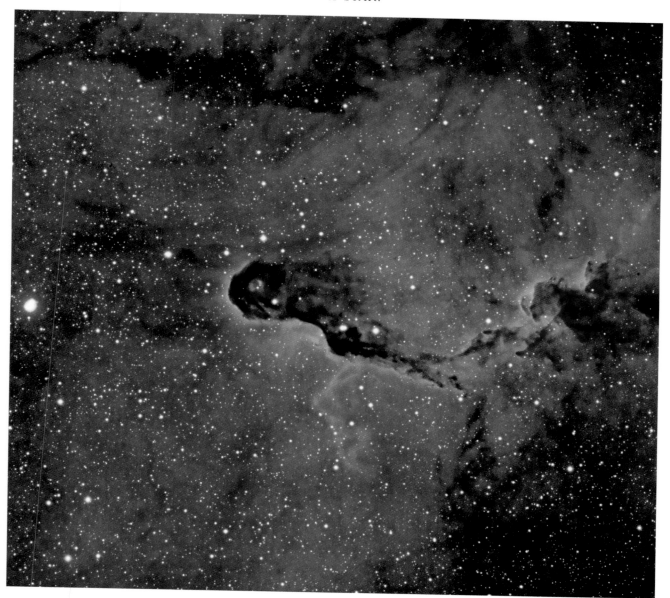

CETUS
The Whale

Located to the south and southeast of Pisces, the constellation Cetus represents the sea monster to which the beautiful Princess Andromeda, daughter of King Cepheus and Queen Cassiopeia of Ethiopia, was to be sacrificed (*see Andromeda*). Perseus saved the day by killing the monster, following which it was placed in the sky alongside the other characters depicted in the legend.

Cetus can be seen in its entirety from virtually anywhere south of latitude 65°N, putting the whole group within reach of observers in central Canada, northern Europe and northern Russia, with at least part of the constellation visible from any location on the planet. The group is best placed for observation during October and November, the star Alrescha in the neighboring constellation Pisces being included on the chart as an aid to identification.

MIRA

Cetus is home to the long period variable star **Mira**, famous for being the first variable to be discovered and the one which gave its name to all other variable stars of its kind. Shining from a distance of around 400 light years, Mira is easily located roughly three-fifths of the way

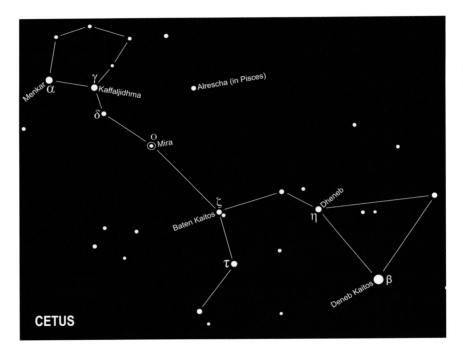

CETUS

from Baten Kaitos to Delta (δ) Ceti as shown on the chart.

On 13 August 1596 the Dutch astronomer David Fabricius observed the

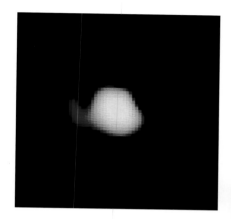

and substantial differences between cycles of variability can occur. Their periods can be anything from less than 100 days to 700 days or more, although successive periods can differ markedly.

As well as being the star after which all variable stars of this type are named, Mira is also the brightest of the long-period variables and only moderate optical aid is needed to follow its complete cycle. Although Mira generally

LEFT: A Hubble Space Telescope image of Mira.

BELOW: The dwarf spiral galaxy NGC 247 in Cetus.

PAGE 82: Another spiral galaxy in Cetus is M77 (NGC 1068).

star and mistook it for a nova (a star that suddenly flares up to several times its original brightness before returning to its original state). In 1603 it was catalogued as a 4th magnitude star by the German astronomer Johann Bayer who entered it as Omicron (o) Ceti in his star atlas *Uranometria*, following which it disappeared from view only to reappear the following year. Subsequent observation finally revealed the true nature of Omicron Ceti. It was the Polish astronomer Johannes Hevelius who named the star Mira, meaning 'wonderful', in his book *Historiola Mirae Stellae* published in 1662 and in which he describes this highly unusual object.

Mira-type variables are long-period variable stars whose brightness oscillates over periods of several months. They are all pulsating red giants, their amplitudes (ranges in brightness from maximum to minimum) averaging out at around 5 or 6 magnitudes, although some are known to vary by as much as 9 or 10 magnitudes. The variations are by no means regular

varies between around magnitude 3 to magnitude 9 or 10 over a period averaging out at 331 days, these values have been known to differ markedly. In 1779 William Herschel recorded that Mira almost attained 1st magnitude, a marked contrast to other times when it has barely reached 4th magnitude. In addition, periods of variability as short as 304 days and as long as 355 days have been recorded.

Mira can often be seen with the naked eye, although binoculars or a small telescope are needed throughout most of its cycle of variability. As is the case with any variable star which remains at or near minimum brightness for extended periods, Mira may well be out of view when you first look for it. However, if you keep watching it will eventually reappear and, once identified, its changes in magnitude will be revealed with continued observation over a period of several weeks. Given its unpredictability, Mira is a star well worth keeping an eye on!

CHAMAELEON
The Chameleon

Located immediately to the south of Musca and Carina, and fairly close to the south celestial pole, the constellation Chamaeleon can only be readily seen from locations on or south of the equator and is completely hidden from view to observers north of latitude 15°N. The chart includes the four main stars in neighboring Musca which can be used as a guide to locating this group.

Chamaeleon is one of the constellations introduced following observations made by Pieter Dirkszoon Keyser and Frederick de Houtman during the 1590s. It contains few stars, all of which are faint and none of which are named. The main stars in the group take the form of a small extended diamond.

THE STARS OF CHAMAELEON

The brightest member of the constellation is **Alpha (α) Chamaeleontis**, a white star shining at magnitude 4.05 from a distance of 64 light years.

Gamma (γ) Chamaeleontis is a magnitude 4.11 red giant, the light from which set off on its journey towards us just over 400 years ago.

Located immediately to the southwest of Alpha is **Theta (θ) Chamaeleontis**, a magnitude 4.34 orange giant star lying at a distance of 155 light years. When seen in binoculars α and θ form a wide pair with a pretty color contrast.

The main stars in Chamaeleon take the form of a small extended diamond, at the opposite end to the Alpha-Theta pairing of which is **Beta (β) Chamaeleontis**, a magnitude 4.24 blue star whose light has taken nearly 300 years to reach us.

Perhaps of most interest to the backyard astronomer is **Delta (δ) Chamaeleontis** which is actually made up of two unrelated stars. The color of magnitude 6.3 blue-white δ¹ contrasts nicely with the orange tint of magnitude 4.45 δ² when viewed through binoculars.

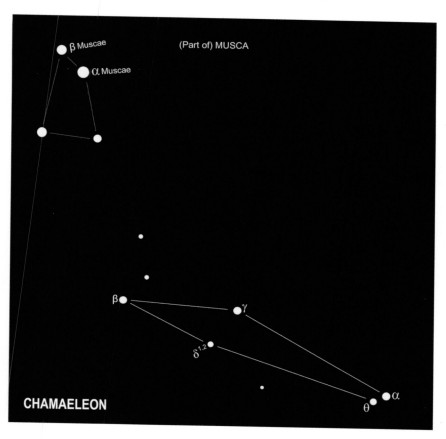

β Muscae
α Muscae
(Part of) MUSCA

β
γ
δ¹,²
α
θ

CHAMAELEON

CIRCINUS
The Compasses

The constellation Circinus takes the form of a tiny elongated triangle located immediately to the west of the two bright stars Alpha (α) and Beta (β) Centauri, both of which are shown on the chart as a guide to locating the group. Circinus can be viewed in its entirety from latitudes south of 30°N making it accessible from southern Mexico, northern Africa, much of India and from locations to the south of these.

The constellation was devised by the French astronomer Nicolas Louis de Lacaille to represent a pair of the dividing compasses used by draftsmen to measure distances. It contains no bright stars and its chief claim to fame seems to be nothing more than the fact that it does actually resemble the object it is supposed to represent, namely a (folded) pair of compasses.

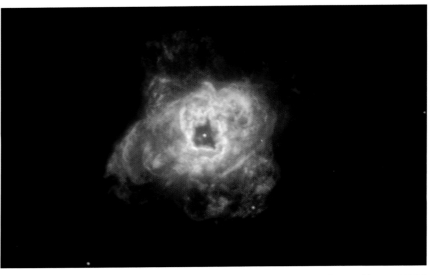

ABOVE: Planetary nebula (see Glossary) NGC 5315 in Circinus.

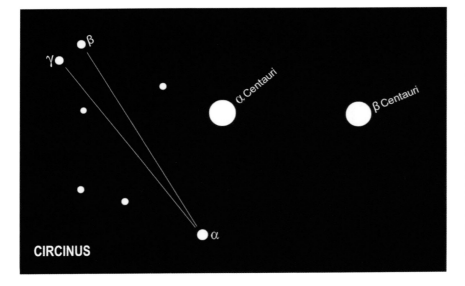

CIRCINUS

THE STARS OF CIRCINUS

Alpha (α) Circini is the brightest star in the group, shining with a magnitude of 3.18 from a distance of 54 light years.

Beta (β) Circini and Gamma (γ) Circini are the two 4th magnitude stars that complete the triangle. Beta has a magnitude of 4.07 and lies at a distance of 100 light years while Gamma is slightly fainter, at magnitude 4.48, and considerably more distant, its light having taken around 440 years to reach us.

COLUMBA
The Dove

Just to the southwest of the star Furud in Canis Major we find the small and somewhat shapeless constellation Columba. Devised in 1592 by the Dutch celestial cartographer Petrus Plancius in order to help fill out this otherwise-empty region of sky, the whole of Columba is visible from latitudes south of 47°N. Furud, together with the nearby brilliant Sirius, also in Canis Major, are depicted on the chart as guides to locating the constellation and, in view of the fact that its two leading stars are reasonably bright, you should have little trouble tracking it down.

THE STARS OF COLUMBA

The brightest star in Columba is the magnitude 2.65 **Phakt (α Columbae)** which shines from a distance of around 260 light years. **Wazn (β Columbae),** located just to the southeast of Phakt, is an orange giant star which, at magnitude 3.12, lies at a distance of 87 light years. The 16th-century Arabic astronomer Al Tizini knew these two stars collectively as **'Al Aghribah'** meaning 'the Ravens'.

Eta (η) Columbae is located on the southern border of Columba, the light from this magnitude 3.96 yellow giant having taken around 500 years to reach us.

Magnitude 3.85 **Delta (δ) Columbae** is another yellow giant, the light from this star reaching us from a distance of 233 light years.

Almost equal in brightness to Delta is **Epsilon (ε) Columbae,** a magnitude 3.86 orange giant star located at a distance of 260 light years.

Gamma (γ) Columbae shines at magnitude 4.36 from a distance of a little over 850 light years.

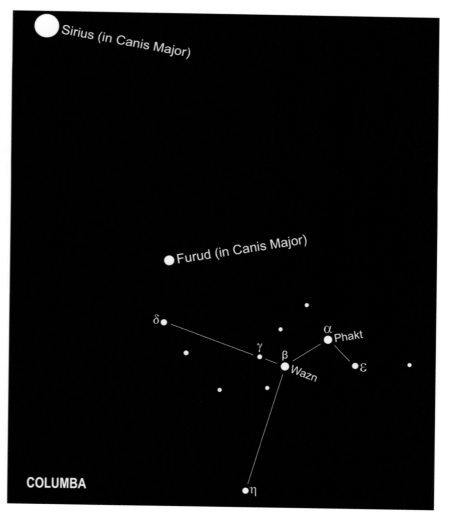

COLUMBA

COMA BERENICES
Berenice's Hair

Located within the asterism of bright stars known as the Diamond of Virgo, comprising Arcturus in Boötes, Spica in Virgo, Denebola in Leo and Cor Caroli in Canes Venatici, the constellation Coma Berenices takes the form of an attractive, scattered collection of faint stars. These are spread out over a small area of sky, the whole of Coma Berenices being visible from latitudes north of 56°S. Although this puts the constellation within the reach of backyard astronomers in Australia, New Zealand and all but the very southern tip of South America, Coma Berenices is best viewed when the sky is really dark and free of moonlight. Observers located south of the equator may have difficulty picking it out at all unless their northern horizon is dark and clear. The brightest stars in Coma Berenices are only of 4th magnitude and seeing this constellation from locations so far south may be regarded as an achievement in itself.

Although this region was observed and recorded by ancient stargazers, the group itself was only made into an official constellation by the Dutch cartographer Gerardus Mercator in the mid-16th century. Prior to this, Arabic astronomers identified this group as the 'Coarse Hair' or 'Tuft' in the tail of the nearby constellation of Leo (the Lion), while the Greek astronomer Eratosthenes identified this swarm of faint stars as the hair of Queen Berenice of Egypt.

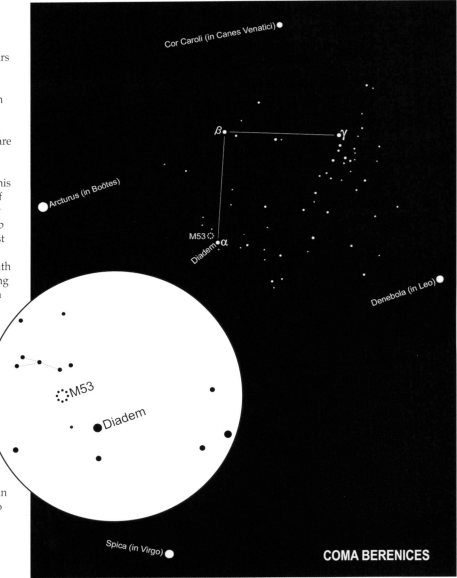

THE STARS OF COMA BERENICES

The three brightest stars in Coma Berenices are all too faint to be easily seen with the naked eye from light-polluted urban areas, and are best sought out with binoculars. These three stars are not actually members of the Coma Star Cluster and are seen superimposed against it. The brightest of the trio is **Beta (β) Comae Berenicis**, a star roughly comparable in size and actual brightness to our own Sun and which shines at magnitude 4.23 from a distance of around 30 light years.

Diadem (α Comae Berenicis) is very slightly fainter, its magnitude 4.32 glow reaching us from a distance of 58 light years.

Gamma (γ) Comae Berenicis is a magnitude 4.35 orange giant whose light has taken around 170 years to reach us.

The legend behind Coma Berenices is quite enchanting and relates to King Ptolemy III of Egypt and his wife Queen Berenice II. While her husband was away from home fighting a war in Asia, Berenice was so concerned for his safety that she offered to sacrifice her hair to the gods if they safeguarded him and allowed him to return home. Ptolemy did indeed come back alive and well, following which happy event Berenice kept her promise to the gods by cutting off her hair and placing it in the temple of Aphrodite at Zephyrium. The gods were so moved by this that they transformed the hair into a constellation, placing it forever among the stars.

The swarm of faint stars that populates the constellation of Coma Berenices is actually an open star cluster. Known as the Coma Star Cluster, it contains around 40 stars which range in brightness between 5th and 10th magnitudes. The Coma Star Cluster lies at a distance of around 250 light years, making it one of the closest objects of its type in the sky. The brightest members are visible to the naked eye and the whole area is rich in stars and well worth sweeping with binoculars.

BELOW: The barred spiral galaxy NGC 4725 in Coma Berenices.

GLOBULAR CLUSTER M53

Coma Berenices plays host to the globular star cluster **Messier 53** (M53) or NGC 5024, discovered by the German astronomer Johann Elert Bode in 1775 and located immediately to the northeast of Diadem. Shining at magnitude 7.6 from a distance of around 58,000 light years, this vast, spherical collection of stars is within the light grasp of a good pair of binoculars which, under clear dark skies, will reveal it as a small, diffuse patch of light.

To see M53 for yourself, first of all locate Diadem then, using the finder chart *(see page 86)* look a short way to the northeast and you should see a tiny dart-shaped pattern of faint stars. These are depicted on the finder chart, which also shows M53 located between these stars and Diadem. As is the case when looking for objects of this type, you need to look for a diffuse patch of luminosity rather than a point of light.

The shimmering glow of Coma Berenices, when viewed under dark and clear skies, impressed the astronomer Joseph Henry Elgie enough for him to write: *'How exquisite is the delicate shimmering of Berenice's fabled hair at this hour! Coma Berenices is, to me, one of the most fairylike objects in the sky. To see its tiny diamond-like stars in their full beauty one should use an opera glass, for then do they appear of a texture that Arachne herself may have spun.'*

RIGHT: A Hubble Space Telescope image of globular cluster M53 (NGC 5024).

CORONA AUSTRALIS
The Southern Crown

Taking the form of a small broken circle of stars, the faint but distinctive Corona Australis lies to the north of Telescopium (the Telescope). Alpha (α) Telescopii, together with the nearby stars Delta[1,2] (δ1,2), Epsilon (ε) and Zeta (ζ) Telescopii are included on the chart for reference. The whole of Corona Australis is observable from the central United States, southern Europe and from latitudes to the south of these.

Corona Australis was one of the 48 constellations drawn up by the Greek astronomer Ptolemy during the second century and is the southern counterpart of Corona Borealis (the Northern Crown). There appear to be no legends directly associated with Corona Australis and the stars forming this group are not particularly prominent although its shape is distinctive in spite of this.

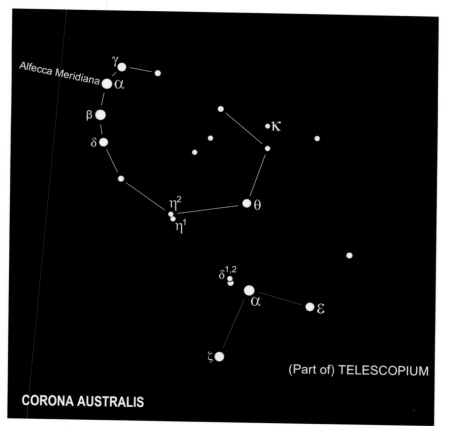

CORONA AUSTRALIS

(Part of) TELESCOPIUM

THE STARS OF CORONA AUSTRALIS

The brightest star in Corona Australis is **Alfecca Meridiana (α Coronae Australis)** which shines at magnitude 4.10 from a distance of 125 light years. The first part of this name comes from the Arabic for 'break', the constellation as a whole taking the form of a broken circle of stars. The second part of the name appears to derive from the Latin for 'south' or 'southern', in all probability to contrast with the similarly-named Alphecca in this constellation's northern counterpart Corona Borealis.

Magnitude 4.11 **Beta (β) Coronae Australis** is an orange giant, the light from which has taken 510 years to reach us.

Gamma (γ) Coronae Australis shines at magnitude 4.23 from a distance of 56 light years.

Immediately to the south of Beta is **Delta (δ) Coronae Australis**, a magnitude 4.57 orange giant whose light has taken around 180 years to reach us.

Yellow giant **Theta (θ) Coronae Australis** shines at magnitude 4.62 from a distance of a little over 550 light years.

DOUBLE STARS IN THE SOUTHERN CROWN

Eta1 (η¹) and Eta2 (η²) Coronae Australis form an optical double, both components of which are white and both of which can be resolved with the naked eye. Magnitude 5.50 Eta¹ shines from a distance of around 350 light years, putting it somewhat closer than Eta² which, at magnitude 5.60, is located at a distance of 500 light years.

The two stars forming the optical double **Kappa (κ) Coronae Australis** are located at vastly differing distances. The light from the brighter magnitude 5.65 component has taken around 1,700 years to reach our planet while that of the fainter magnitude 6.30 star set off on its journey towards us around 490 years ago. Both components can be resolved in a small telescope.

BELOW: A star field and region of reflection nebulosity in Corona Australis.

CORONA BOREALIS
The Northern Crown

Located immediately to the east of the constellation Boötes, and taking the form of a distinctive semicircle of reasonably bright stars, Corona Borealis lies only a short way north of the celestial equator. Consequently, the entire constellation can be seen from anywhere north of latitude 50°S, making it a potential target for observers in Australia, New Zealand, South Africa and virtually the whole of South America. The two bright stars Arcturus and Izar, both in Boötes, are shown on the chart to help you identify the group.

Corona Borealis is one of the limited number of constellations that resemble the objects they are supposed to depict, in this case the crown presented by Bacchus to Ariadne, daughter of King Minos of Crete. According to Greek mythology the crown was made by the supreme goldsmith Hephaestus at his underwater smithy and given to Ariadne after she had been deserted by the unfaithful Theseus. After Ariadne's death, the gods placed her crown in the sky. Alternative stories include that of Arabic astronomers who identified the group as a broken plate (presumably because the stars of the constellation don't form a complete circle).

DOUBLE STARS IN THE NORTHERN CROWN

Sigma (σ) Coronae Borealis is a binary star in which the two components are actually orbiting each other. Hovering on the edge of naked-eye visibility, Sigma lies a little way to the northeast of Iota and, once located, its magnitude 5.7 and 6.7 yellowish components are easily resolved through a small telescope. Another binary for small telescopes is **Zeta (ζ) Coronae Borealis**, a system containing two blue-white companion stars of magnitudes 5.1 and 6.0 located at a distance of 470 light years.

Located just to the east of Sigma is the optical double star **Nu (ν) Coronae Borealis**, the two components of which lie at different distances. The brightest is a magnitude 5.20 red giant shining from a distance of around 640 light years, somewhat further away than its slightly fainter companion, an orange giant star with a magnitude of 5.40, the light from which has taken around 590 years to reach us. The components of Nu are far enough apart to be resolvable in binoculars.

THE STARS OF CORONA BOREALIS

Alphecca (α Coronae Borealis) is the brightest star in Corona Borealis, shining at magnitude 2.22 from a distance of around 75 light years. Alphecca derives its name from the Arabic for 'break', a possible allusion to the constellation itself being an incomplete circle of stars. The star Alfecca Meridiana in Corona Australis, the southern counterpart of Corona Borealis, derives its name in a similar way.

Nusakan (β Coronae Borealis) shines at magnitude 3.66 from a distance of around 115 light years.

The light from magnitude 3.81 **Gamma (γ) Coronae Borealis** set off towards us around 150 years ago.

Theta (θ) Coronae Borealis is a blue giant, the magnitude 4.14 glow of which reaches us from a distance of around 375 light years.

The white giant star **Iota (ι) Coronae Borealis** has a magnitude of 4.98 and lies at a distance of a little over 300 light years.

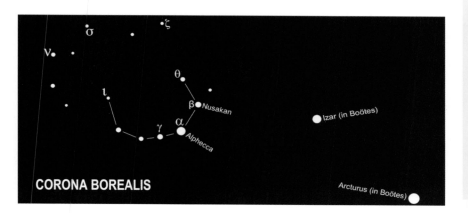

CORONA BOREALIS

CORVUS & CRATER
The Crow & The Cup

The stars forming Corvus and Crater are not particularly bright, although the shapes of these two groups are quite distinctive. Both constellations lie in the area of sky to the south and southwest of the bright star Spica in Virgo which is depicted on the chart for guidance. Both Corvus and Crater lie a little to the south of the celestial equator, resulting in both constellations being visible from every inhabited part of the world lying south of latitude 65°N.

Both Corvus and Crater are associated with the legend involving Apollo and his request that the crow fetch him a cup of the water of life *(see Hydra)* which resulted in the crow and the cup being banished to the sky. Both constellations lie immediately to the north of the long and meandering constellation Hydra (the Water Snake) who guards the cup and its contents from the crow, the outcome of which is that the unfortunate bird is condemned to eternal thirst.

THE STARS OF CORVUS

Gienah (γ Corvi) is the brightest star in Corvus, shining at magnitude 2.58 from a distance of 154 light years. This star derives its name from the Arabic *'janah'* meaning 'wing' and, although here it represents the wing of a crow, it also represents the outstretched wing of the constellation Cygnus (see Cygnus).

The light from magnitude 2.65 yellow giant star **Kraz (β Corvi)** has taken 140 years to reach us.

Magnitude 2.94 **Algorab (δ Corvi)**, the name of which comes from the Arabic *'janah al-ghurab'* meaning 'the Raven's Wing', shines from a distance of 87 light years.

The orange giant **Minkar (ε Corvi)** completes the distinctive quadrilateral shape of Corvus, the magnitude 3.02 glow from this star reaching us from a distance of around 320 light years.

The light from magnitude 4.02 **Alchita (α Corvi)** set off towards us 49 years ago.

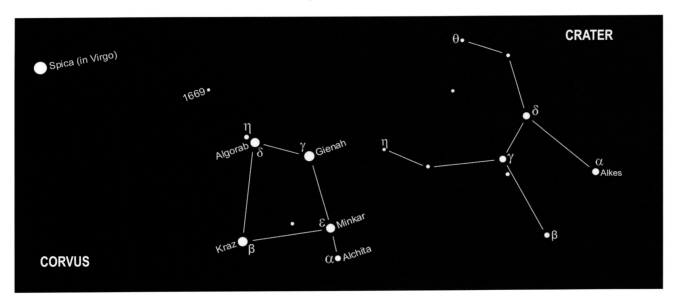

THE STARS OF CRATER

Although the general shape of Corvus may not resemble that of a crow, its close neighbor Crater does bear an overall likeness to a cup complete with base. The stars forming Crater are generally fainter than those in Corvus, the brightest member of which is the yellow giant **Delta (δ) Crateris** which shines at magnitude 3.56 from a distance of around 180 light years.

Gamma (γ) Crateris glows at magnitude 4.06, its light having taken 82 years to reach us.

Another yellow giant is **Alkes (α Crateris)**, a star which derives its name from the Arabic **'al-ka's'** meaning 'the (Wine) Cup'. The magnitude 4.08 glow of Alkes set off on its journey towards us around 160 years ago.

Completing the base of the cup is magnitude 4.46 **Beta (β) Crateris**, a white star shining from a distance of around 340 light years.

The rim of the cup is marked by **Eta (η)** and **Theta (θ) Crateris**. Eta is a white giant star shining at magnitude 5.17 from a distance of 250 light years, slightly fainter, albeit a little closer, than magnitude 4.70 Theta, the light from which set off towards us 280 years ago.

DOUBLE STARS IN CORVUS

Corvus contains several double stars worth checking out including the optical double formed from Algorab and magnitude 4.30 **Eta (η) Corvi** which, at a distance of 60 light years, lies somewhat closer to us than Algorab. The pair form a pretty sight when viewed through binoculars or a small telescope, the latter of which will reveal that Algorab is itself a double with a nearby 9th magnitude bluish companion star.

Located to the northeast of Algorab and Eta is the double star Struve 1669, its name derived from its number in the catalogue of double stars compiled by the German astronomer Friedrich Georg Wilhelm von Struve and his son Otto Wilhelm von Struve during the 19th century. The pair of 6th magnitude yellowish stars that form Struve 1669 are an attractive sight when viewed through small telescopes.

RIGHT: The Antennae Galaxies (NGC 4038 and NGC 4039) are a pair of interacting galaxies in Corvus.

CRUX
The Cross

Although Crux is the smallest of the constellations, it is also one of the most famous, descriptions of the stars forming this particular group having been recorded as far back as the early 16th century. It was in 1516 that the Italian navigator Andreas Corsali first described Crux as a separate constellation, although in 1504 his fellow navigator Amerigo Vespucci had drawn attention to a number of important southern stars. Amongst their number was a particular group of six which were probably the four main stars in Crux together with the nearby Alpha and Beta Centauri *(see Centaurus).*

The inclusion of Alpha and Beta Centauri in Vespucci's records is perhaps appropriate in that the stars forming Crux were once part of the neighboring constellation Centaurus (the Centaur), a group included by Ptolemy on the list of 48 constellations he drew up during the second century. Crux first appeared as a individual constellation on celestial globes made during the 1590s from which time its identity as a separate group was assured. A regular feature on atlases since then, modern star charts depict it surrounded on three sides by Centaurus.

Crux lies not far from the south celestial pole, the consequence of which is that it can never be seen from mid-northern latitudes. The constellation is visible in its entirety from locations south of latitude 25°N, putting it within the reach of observers in Central America, southern India and anywhere further south. When viewed from places on or south of the equator Crux is best placed for viewing around April when it is located fairly high up in the sky. Crux is set against the backdrop of the Milky Way and the area is well worth sweeping with binoculars which will reveal some wonderful star fields. When seen through binoculars, the orange/red tint of Gamma Crucis contrasts with the blue or blue-

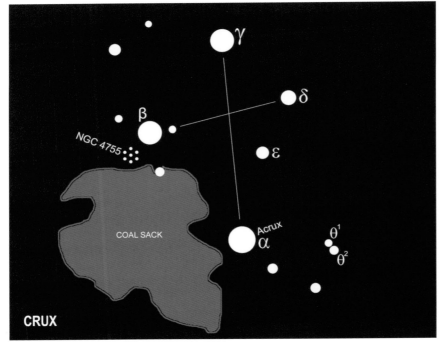

ABOVE: The Jewel Box (NGC 4755) open star cluster.

OPPOSITE: The Coal Sack dark nebula in Crux.

THE STARS OF CRUX

Acrux (α Crucis) is the brightest star in the group, shining at magnitude 0.77 from a distance of around 320 light years. Deriving its name from a combination of the full Bayer reference (Alpha Crucis), Acrux is actually a double star, its individual magnitude 1.4 and 1.9 components resolvable in small telescopes.

The light from magnitude 1.25 **Beta (β) Crucis** has taken around 280 years to reach us.

The blue giant **Delta (δ) Crucis** shines at magnitude 2.79 from a distance of just under 350 light years.

The orange giant star **Epsilon (ε) Crucis** offsets the symmetry of the Cross, its magnitude 3.59 glow reaching us from a distance of 230 light years.

The red giant **Gamma (γ) Crucis** is another double star, although the association between the two components is nothing more than a line of sight effect. Gamma shines at magnitude 1.6 and, when viewed through binoculars or a small telescope, seems to be accompanied by a fainter magnitude 6.7 companion. However, Gamma lies at a distance of 89 light years, only a quarter of the distance of the fainter component.

When seen against the backdrop of star fields against which they lie, **Theta¹ (θ¹)** and **Theta² (θ²) Crucis** form a pretty pair when viewed through binoculars. Magnitude 4.32 Theta¹ lies at a distance of around 230 light years, a little over a quarter of the distance of magnitude 4.72 Theta² which shines from around 800 light years.

astronomer Nicolas Louis de Lacaille, it can be resolved even through binoculars which will show it to contain a number of bright bluish-white stars. It also plays host to one supergiant star with a distinctly red tint and which, when viewed through telescopes, lends a beautiful splash of color to the cluster.

NGC 4755 is also known as the Jewel Box, a name coined by the English astronomer John Herschel who described the cluster as: ' . . . *an extremely brilliant and beautiful object when viewed through an instrument of sufficient aperture to show distinctly the very different colour of its constituent stars, which give it the effect of a superb piece of fancy jewellery.*' Located at a distance of around 6,500 light years, and containing over 100 individual stars, the Jewel Box cluster is a must-see object for the backyard astronomer.

white hues of the other three stars making up the main form of this tiny but interesting constellation.

The cruciform shape of Crux should not be confused with the asterism of four bright stars which lie across the border between the nearby constellations Carina and Vela. Known collectively as the False Cross (*see Carina*), these are found a little way to the west of Crux and form a larger but very similar pattern.

THE JEWEL BOX CLUSTER

Visible to the naked eye as a 4th-magnitude fuzzy star-like object immediately southeast of Beta Crucis, the open star cluster NGC 4755 is a splendid object. Discovered by the French

THE COAL SACK

Located at a distance of around 500 light years, the Coal Sack dark nebula takes on the appearance of a dark patch silhouetted against the brighter background of the Milky Way. Visible to the naked eye, its overall appearance is offset slightly by the presence of a few faint foreground stars. The Jewel Box cluster is seen immediately to the north of the Coal Sack, although this association is no more than a line of sight effect. The Coal Sack is the most prominent object of its type and, as its name suggests, appears as a huge, dark cloud of absorbing dust blotting out the light from the stars beyond. Most of the Coal Sack lies in Crux although it overlaps slightly into the neighboring constellations Centaurus and Musca.

CYGNUS
The Swan

During September evenings, when viewed from mid-northern latitudes, the prominent constellation Cygnus is situated at or near the overhead point and is one of the leading star groups of northern summer. Observers south of the equator will see its distinctive shape perched above the northern horizon, attended by the brilliant Vega in Lyra to its west and Altair, the leading star in Aquila, higher up and to the south of this pair. Cygnus can be seen in its entirety from latitudes north of 29°S with portions of Cygnus visible from anywhere to the north of latitude 62°S. The chart depicts Cygnus together with the two nearby bright stars Altair (in Aquila) and Vega (in Lyra) which, together with Deneb, make up the conspicuous asterism known as the Summer Triangle.

Ancient astronomers have likened this group to a whole range of different birds, their ranks including a hen, a flying eagle, a partridge and a pigeon. The name we now apply to the constellation originated with the Romans who linked it with the mythical swan identified with Cycnus, the son of Mars. Cygnus does indeed conjure up the image of a huge celestial bird superimposed against the Milky Way and flying southwards across the sky. An old Polish hymn talks of *'a sleeping swan's white plumage fringed with gold'*, a clear reference to the glow of the Milky Way that surrounds the heavenly bird, the overall effect of which can be striking on clear, dark, moonless nights.

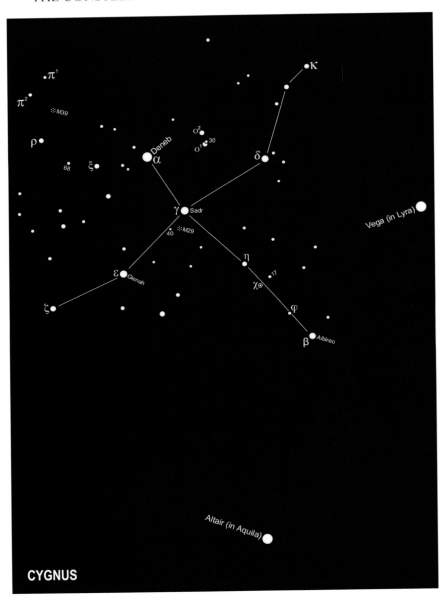

CYGNUS

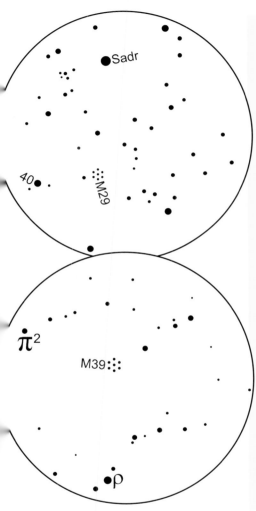

THE STARS OF CYGNUS

Deneb (α Cygni) is the brightest star in Cygnus. Deriving its name from the Arabic *'dhanab al-dajaja'* meaning 'the Hen's Tail', this magnitude 1.25 white supergiant shines from a distance of around 1,400 light years. Depicting the tail of Cygnus, Deneb marks the northern end of the constellation and from here the swan's outstretched wings and characteristic long neck can be picked out.

Magnitude 2.23 **Sadr (γ Cygni)** is another white supergiant, the light from which has taken around 1,800 years to reach us. This star takes its name from the Arabic *'sadr al-dajaja'* meaning 'the Hen's Breast'.

Shining at magnitude 2.48 from a distance of 73 light years **Gienah (ε Cygni)** is an orange supergiant star which derives its name from the Arabic *'janah'* meaning 'wing', as does the star in Corvus which bears the same name *(see Corvus)*.

Delta (δ) Cygni is a white star of magnitude 2.86 shining from around 170 light years.

The yellow supergiants **Zeta (ζ) and Kappa (κ) Cygni** mark the tips of the swan's wings. Zeta shines at magnitude 3.21 from a distance of 143 light years, putting it slightly further away than magnitude 3.80 Kappa, the light from which has taken just 124 years to reach us.

Albireo (β Cygni) depicts the beak of Cygnus and appears as a 3rd magnitude star when viewed with the naked eye. Closer inspection will reveal the true beauty of this star which is held to be one of the most attractive double stars in the sky and a showpiece for small telescopes.

The brightest component has a yellowish tint and, shining at magnitude 3.05 from a distance of around 430 light years, contrasts beautifully with its bluish magnitude 5.10 companion, the light from which has taken just over 400 years to reach us. Albireo has long been held in high esteem by stargazers, including Joseph Henry Elgie who said: '*Of Beta Cygni (Albireo), its chief interest consists in it being a very beautiful double. The component star . . . is an easy object in a small telescope. Webb . . . describes Albireo as topaz yellow in color and the component as sapphire blue. He thinks the double to be one of the finest in the heavens . . .'*.

Perhaps slightly less impressive than Albireo is the trio of stars **Omicron[1] (o[1])**, **Omicron[2] (o[2])** and **30 Cygni** which nonetheless form a pretty sight in binoculars. Both Omicron[1] and Omicron[2] are orange supergiants, the latter shining at magnitude 3.96 from a distance of a little over 1,000 light years. Although estimates as to the distance of magnitude 3.80 Omicron[1] vary, the generally accepted value is around 800 light years. The magnitude 4.80 white star 30 Cygni, which can be seen lying close to Omicron[1], shines from a distance of around 600 light years and is located in virtually the same line of sight as seen from our location in space.

The distinctive shape of Cygnus has led to the constellation being known as the Northern Cross, echoed by the American astronomer Percival Lowell who wrote in 1844 that the countless splendors in the sky were *'crowned by the blazing Cross hung high o'er all'*.

OPEN STAR CLUSTER M39

The initial discovery of the open star cluster **Messier 39** (M39) or NGC 7092, located a short way to the northeast of Deneb, may have been made by the rather grandly named French astronomer Guillaume-Joséph-Hyacinthe-Jean-Baptiste Le Gentil de la Galaisière (usually known as Le Gentil) who, in 1750, noted: *'At the tip of the tail of Cygnus . . . a large cloud, bigger at one end than the other . . . can be seen without the telescope.'* However, the discovery is usually credited to Charles Messier who, in 1764, described this object as a: *'Cluster of stars near the tail of Cygnus . . . '*.

Located at a distance of 825 light years and shining with an overall magnitude of

around 5.5, M39 is a large but fairly loose gathering of stars and can just be made out with the naked eye under really dark and clear skies. Because of its large apparent size, M39 is best observed either in the wide field of view of binoculars or through telescopes using a low magnification, both of which should reveal the overall triangular shape of the cluster.

You can track M39 down by using binoculars to follow the line of faint stars **Xi (ξ)**, **68**, **Rho (ρ)**, **Pi² (π²)** and **Pi1 (π¹)** away from Deneb. The cluster will be seen to form a small triangle with Rho and Pi² as depicted on the finder chart *(see page 97)*. Binoculars will reveal some of the individual stars in the cluster and even a small telescope will bring out a bright handful of member stars.

OPEN STAR CLUSTER M29

Discovered by Charles Messier in 1764, the open star cluster **Messier 29** (M29) or NGC 6913 shines with an overall magnitude of 7.1 from a distance of around 4,000 light years. Containing

around 50 member stars, M29 is situated immediately to the southeast of the star Sadr within an area rich in stars and is a fairly easy target for the backyard stargazer. M29 can be found by first of all locating Sadr together with the nearby guide star 40 Cygni. You can then star hop your way to the cluster through the field of faint stars depicted on the finder chart *(see page 97)*. M29 is a little fainter than M39 and some patience may be required during the search. Once the cluster is located, binoculars will reveal it as a distinct fuzzy patch of light, with small telescopes bringing out a number of individual cluster members.

LONG PERIOD VARIABLE STAR CHI CYGNI

Located in the neck of the swan, immediately to the southeast of 17 Cygni, and a little over half way between **Phi (φ) and Eta (η) Cygni**, is the Mira-type variable star **Chi (χ) Cygni**. As far as stars of this type are concerned, Chi Cygni is unusual in that it varies between 5th and 13th magnitude. This is one of the largest variations in brightness known, the complete cycle of variability taking place over a period of 407 days. What is even more unusual is that Chi Cygni can be

ABOVE LEFT: Open star cluster M39 (NGC 7092).

LEFT: Open star cluster M29 (NGC 6913).

OPPOSITE: A section of the Veil Nebula, which is part of the Cygnus Loop, the large but relatively faint remnant of a star which exploded over 5,000 years ago.

anything up to two magnitudes brighter when at maximum, with peaks in magnitude of 4.6 and even 3.3 having been recorded.

Because Chi Cygni is so faint when at or near minimum brightness, you may have problems picking it out from the surrounding field of faint stars, in spite of it having the distinct reddish tint typical of stars of this type. The best way to look for Chi Cygni is to keep an eye on its location and wait for it to appear during its approach to maximum brightness. The reddish hue of Chi Cygni should help you identify it and, once located, comparing its brightness to nearby stars will reveal its subsequent changes in magnitude. You can follow the star with binoculars or a small telescope through most, but not all, of its entire period of variability, although it should be borne in mind that variable stars of this type take a long time to complete a full cycle. As a consequence, observations you start now may not be completed until the following year!

DELPHINUS
The Dolphin

Delphinus is a small but very distinctive diamond-shaped pattern of stars which can be viewed in its entirety from all inhabited parts of the world, located a little way to the east of the trio of bright stars Alshain, Altair and Tarazed in the neighboring constellation Aquila.

Greek legend associates Delphinus with the talented poet and musician Arion, whose skills playing the lyre were unequalled. Arion was returning to Greece by boat from Sicily although the sailors on board the ship plotted to kill him and steal his wealth. They sailors did show some compassion, allowing Arion to play one of his favorite songs before he died, although

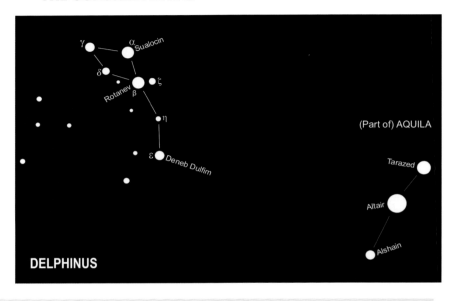

DELPHINUS

THE STARS OF DELPHINUS

Shining at magnitude 3.64, **Rotanev (β Delphini)** is the brightest star in Delphinus. Located at a distance of a little over 100 light years, this star is somewhat closer than magnitude 3.77 **Sualocin (α Delphini)**, the light from which has taken around 250 years to reach us. Under clear skies, binoculars will reveal the color difference between these two stars, the blue-white tint of Sualocin contrasting with the yellowish hue of Rotanev.

An interesting story surrounds the origin of the names of the above two stars. When spelled backwards, Sualocin and Rotanev read as Nicolaus Venator, this being the Latinized version of Niccolo Cacciatore, who happens to have been assistant to Giuseppe Piazzi, the director of the Palermo

Observatory in Sicily during the early 19th century. Cacciatore succeeded Piazzi as director and seems to have called these two stars after himself, the names first appearing in the observatory's 1814 star catalogue. Although a rather unusual origin for star names, the astronomer responsible is indeed immortalized amongst the stars!

Gamma (γ) Delphini is a pretty double star shining from a distance of around 100 light years. Its magnitude 4.27 and 5.15 components can be resolved in a small telescope which should reveal the yellowish tints of both stars.

Eta (η) Delphini shines at magnitude 5.39 from around 235 light years.

Deneb Dulfim (ε Delphini) shines at magnitude 4.03 from a distance of around 350 light years, the name of this star being derived from the Arabic for 'Tail of the Dolphin'. Deneb is a common star name and occurs in many other constellations in one form or another, including Deneb Kaitos in Cetus, Deneb in Cygnus and Deneb Algiedi in Capricornus.

The light from magnitude 4.43 **Delta (δ) Delphini** reaches us from a distance of around 220 light years, putting it at the same distance as magnitude 4.64 **Zeta (ζ) Delphini**, the star which completes the main outline of Delphinus.

before he had finished they noticed that a school of dolphins had been attracted to the ship by Arion's music. The terrified attackers quickly threw Arion overboard, although he was lucky enough to land on the back of one of the dolphins and was eventually carried to the safety of land. As a reward, Apollo placed the dolphin up among the stars where he can be seen to this day.

The constellation Delphinus is seen here set against a backdrop of rich star fields.

DORADO
The Goldfish

Dorado is another of the constellations that were introduced by Pieter Dirkszoon Keyser and Frederick de Houtman in the 1590s, appearing on the celestial globe produced by Petrus Plancius in 1598 and can be located by extending an imaginary line from the prominent star Canopus in Carina through a point roughly midway between Beta (β) and Gamma (γ) Pictoris. Taking this line roughly twice as far again will lead you to Alpha (α) Doradus from where the rest of the group can be picked out. Dorado is visible in its entirety from locations south of latitude 20°N, the star Canopus, together with the main outline of Pictor, included on the chart as a guide to locating the constellation.

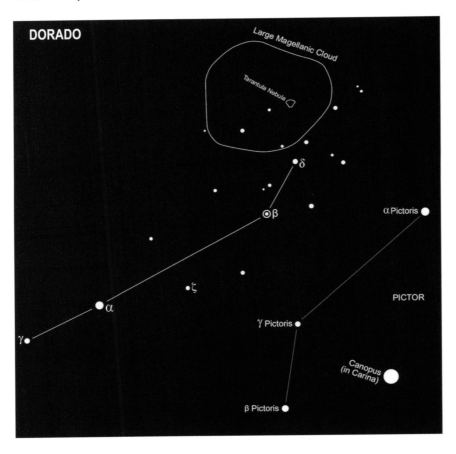

THE STARS OF DORADO

Alpha (α) Doradus is the brightest star in the constellation, shining at magnitude 3.30 from a distance of around 170 light years,

Located at a distance of around 1,000 light years, **Beta (β) Doradus** is a Cepheid variable (see Cepheus), its magnitude ranging between around 3.40 to 4.10 over a period of 9.84 days. You can compare its brightness with the nearby magnitude 4.71 **Zeta (ζ) Doradus** and magnitude 4.34 **Delta (δ) Doradus**, located to either side of Beta.

Gamma (γ) Doradus shines at magnitude 4.26 from a distance of 67 light years and completes the line of stars that forms Dorado.

THE LARGE MAGELLANIC CLOUD

Dorado plays host to the Large Magellanic Cloud (LMC) which straddles the border of Dorado and the adjoining constellation Mensa. Like its counterpart the Small Magellanic Cloud (see Tucana), the LMC is a dwarf irregular galaxy and a member of the Local Group of Galaxies (see Glossary). Located at a distance of over 160,000 light years, the LMC is a sizeable system measuring around 14,000 light years across. Resembling a detached portion of the Milky Way, it is visible even in moonlit skies and is plainly visible to the unaided eye. Time spent sweeping the LMC with binoculars or a telescope will be well rewarded.

.. *but fainter.'* When we gaze at this magnificent object we are looking at one of the most active regions of star formation known, the heart of the Tarantula Nebula containing a massive recently-formed star cluster measuring some 35 light years in diameter. The stars within this cluster are producing prodigious amounts of energy which cause the gas in the Tarantula Nebula to shine.

LEFT: The Large Magellanic Cloud.

BELOW: The Tarantula Nebula (NGC 2070).

THE TARANTULA NEBULA

The Tarantula Nebula (NGC 2070) is the brightest part of the entire Large Magellanic Cloud and by far the largest-known diffuse nebula within the Local Group of Galaxies. Its name arises from the fact that its spidery outer filaments and streamers, stretching away to cover an area of space some 1,000 light years across, resemble the legs of a gigantic tarantula. This highly luminous object can be identified with the naked eye at a distance of 160,000 light years, bearing testimony to its colossal size and brilliance. Indeed, so luminous is the Tarantula Nebula that if it were as close to our planet as the Orion Nebula it would cast shadows.

Originally thought to be a star, the true nature of the Tarantula Nebula was determined by the French astronomer Nicolas Louis de Lacaille who described it as: '*(Like the nucleus of a fairly bright comet)* .

DRACO
The Dragon

The long and winding form of Draco can be found curving its way around Polaris, the Pole Star, and its host constellation Ursa Minor. Because of its far-northern location, only those observers at latitudes north of 4°S can see the entire group, when it is suitably placed in the sky, and because only parts of the constellation are accessible from latitudes north of 42°S, it is effectively lost to view for observers in the southern hemisphere. The best time to view Draco is in July and August, at which time it can be found at or near the overhead point when viewed from mid-northern latitudes.

Although Draco is one of the largest constellations, its leading star Etamin is only of 2nd magnitude and the constellation as a whole isn't particularly prominent. However, it can be picked out reasonably easily by using the stars of Ursa Minor as a guide. The quadrilateral of stars forming the 'head' of Draco, including Etamin, can be found just to the north of the brilliant star Vega in the constellation Lyra. Vega, together with the main stars in Ursa Minor, are included on the chart as a guide to locating Draco. Provided the sky is dark and clear you should have little difficulty in tracing out this meandering constellation as it winds its way around the northern sky.

One of the 12 labours set for Hercules was to steal some of the fruit from the golden apple tree which had been given to Hera as a wedding gift following her marriage to Zeus. The tree was planted on the slopes of Mount Atlas and was

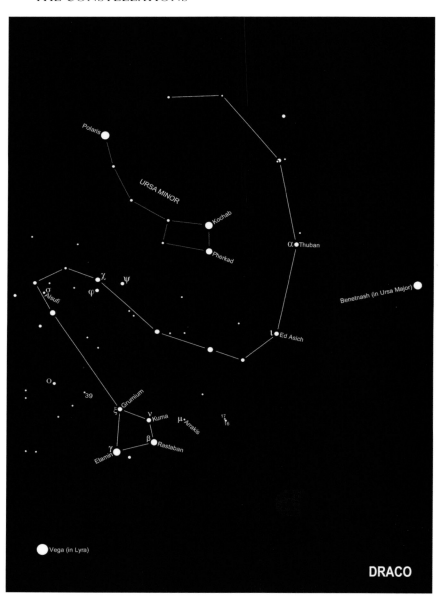

DRACO

WHAT STAR?

guarded by the Hesperides, the daughters of Atlas. As guardians, however, they proved to be somewhat untrustworthy. They kept picking and eating the apples themselves, so a replacement guard was installed in the form of a dragon named Ladon. According to legend, Ladon had

100 heads and presented a rather daunting sight to onlookers. Hercules was undeterred and promptly killed the dragon with poisoned arrows before making off with several apples for himself! Hera placed the unfortunate dragon in the sky where it can be seen

to this day in the form of the constellation Draco.

The trio of galaxies NGC 5985, NGC 5982 and NGC 5981 in Draco.

THE STARS OF DRACO

The position of the north celestial pole is currently marked by the star Polaris in Ursa Minor, although this has not always been the case. Because the Earth's axis of rotation is 'wobbling', the positions of the celestial poles are slowly changing. This wobbling motion is known as precession *(see Glossary)*, a consequence of which is that, at around the time the Egyptians were building the Pyramids, the north celestial pole was marked by the star **Thuban (α Draconis)**. To put it another way, Thuban was the 'Pole Star' of the Ancient Egyptians. Deriving its name from the Arabic *'ra's al-tinnin'* meaning 'the Serpents Head', Thuban shines at magnitude 3.67 from a distance of just over 300 light years.

Etamin (γ Draconis) is the brightest star in Draco and one of the four stars forming the head of the Dragon. The light from this magnitude 2.24 orange giant reaches us from a distance of 154 light years.

Magnitude 2.79 **Rastaban (β Draconis)** is a yellow supergiant star located at a distance of 380 light years.

One of the most easily resolved double stars in the northern skies, and the faintest of the four stars forming the head of Draco, is **Kuma (ν Draconis)**. Shining from a distance of 99 light years, and with a combined magnitude of 4.13, the individual magnitudes of the two stars forming the Kuma system are 4.86 and 4.89, making them almost identical in brightness. Those with really keen eyesight may be able to split this pair with the naked eye, and binoculars bring them out well.

The orange giant **Grumium (ξ Draconis)**, which shines at magnitude 3.73 from a distance of 113 light years, completes the outline of the head of Draco.

The magnitude 4.97 **Arrakis (μ Draconis)** can be found close to the Dragon's head and derives its name from the Arabic *'al-raqis'* meaning 'the Trotting Camel'. The light from Arrakis has taken around 90 years to reach our planet.

Magnitude 4.67 **Alsufi (σ Draconis)** shines from a distance of just 18.7 light years, making it one of the closest naked-eye stars.

Ed Asich (ι Draconis) is an orange giant star with a magnitude 3.29 glow reaching us from a distance of 101 light years. Ed Asich takes its name from the Arabic *'al-dhikh'* meaning 'the Male Hyena'.

MORE DOUBLE STARS IN DRACO

The double star Kuma *(see above)* can be used to locate **Omicron (ο) Draconis**, another double which can be resolved in binoculars. The brighter component shines at magnitude 4.7 and has an orange tint which contrasts with the blue of its magnitude 7.5 companion. Omicron can be located by extending the line taken from Kuma, through Grumium and 39 Draconis.

Located near the middle of the Dragon's back, and forming a tiny but distinctive triangle with the nearby **Chi (χ)** and **Phi (φ)**, is the double star **Psi (ψ)** Draconis. The individual components are easily resolved in binoculars although you may need a small telescope to bring out the yellowish and bluish tints of the magnitude 4.6 and 5.8 stars forming this double.

Draco is home to yet another double star resolvable in binoculars, this being the one formed from the two stars 16 and 17 Draconis, a widely-spaced pair which can be tracked down by following a line from Kuma, through Arrakis and on roughly as far again. The closer of the two stars is 17 Draconis which, at magnitude 5.07, shines from a distance of around 410 light years. Slightly further away is 16 Draconis, the magnitude 5.53 glow of which light reaches us from a distance of around 425 light years.

EQUULEUS
The Little Horse

Introduced by Ptolemy on his list of 48 constellations drawn up in the second century AD and located immediately to the west of Pegasus is the constellation Equuleus. Taking the form of a tiny trapezium of faint stars, Equuleus is visible in its entirety from anywhere north of latitude 77°S, making it accessible to observers in almost any inhabited part of the world.

Equuleus can be made out with the naked eye under clear and dark skies, the comparatively bright star Enif in neighboring Pegasus being a useful guide. First of all locate Enif, with the help of binoculars if required, and then look for the triangle formed from the three stars Kitalpha, Gamma Equulei and Delta Equulei. Once these have been identified the rest of this faint constellation can be picked out.

THE STARS OF EQUULEUS

Magnitude 3.92 **Kitalpha (α Equulei)** is the brightest star in the constellation, its name being derived from the Arabic **'qit at al-faras'** meaning 'the Section of the Horse'. Kitalpha lies at a distance of around 190 light years.

Delta (δ) Equulei lies much closer to us, the light from this magnitude 4.47 star having taken just 60 years to reach our planet.

Beta (β) Equulei shines at magnitude 5.16 from a distance of 330 light years.

Located at a distance of 175 light years, and completing the main outline of Equuleus, is magnitude 5.30 **Epsilon (ε) Equulei.**

The white giant star **Gamma (γ) Equulei** shines at magnitude 4.70, its light having set off towards us 118 years ago. Gamma forms a wide double with the magnitude 6.07 star **6 Equulei**. However, the relationship between these two stars is nothing more than a line of sight effect, 6 Equulei having been located at a distance of around 440 light years, nearly four times that of Gamma. Provided the sky is really dark, clear and moonless, you may just be able to pick out the faint glow from 6 Equulei and so resolve this pair with the naked eye, although you will probably need the help of a pair of binoculars, which will bring out both stars quite well.

Enif (in Pegasus)

δ

γ

6

β

Kitalpha α

ε

EQUULEUS

ERIDANUS
The River

With portions of this long and winding constellation being visible from almost anywhere in the world, Eridanus can be viewed in its entirety from all latitudes south of 32°N, extending as it does from the region of sky close to the brilliant Rigel in Orion. From here it flows is a westerly direction before turning and making its way southwards, ending in the glow of its leading star Achernar deep in the southern hemisphere.

According to legend Phaeton, the son of Helios, the god of the Sun, asked his father to let him drive the chariot of the Sun across the sky. Helios was eventually persuaded that this was a good idea and allowed Phaeton to set off, although this was to be an ill-fated journey due to the fact that Phaeton lost control of the steeds pulling the chariot. Zeus was watching events unfold and, to ensure that a disaster was averted, hurled a thunderbolt at the chariot which, along with the hapless Phaeton, ended up plunging into the River Eridanus far below.

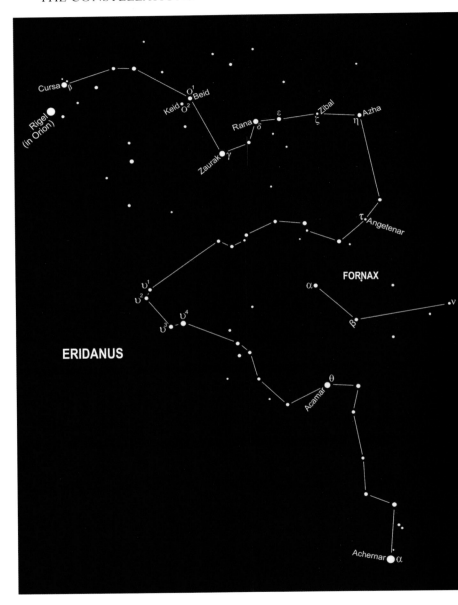

THE STARS OF ERIDANUS

Achernar (α Eridani) marks the southern extremity of the River Eridanus and is the brightest star in the constellation. Its name is derived from the Arabic *'akhir al-nahr'* meaning 'the River's End', an appropriate title bearing in mind its location. Achernar shines with a magnitude of 0.45 from a distance of 140 light years.

Cursa (β Eridani) is the second brightest star in the constellation. Shining from a distance of 89 light years, it marks the northern extremity of Eridanus and is easily located just to the northwest of the brilliant Rigel, one of the leading stars in the neighboring constellation Orion. Magnitude 2.78 Cursa derives its name from the Arabic for 'The Foremost Footstool [of Orion]', a befitting title bearing in mind that Rigel depicts the left foot of Orion.

Acamar (θ Eridani) is slightly fainter than Cursa, shining with a magnitude of 2.88 from a distance of around 160 light years. Acamar marks the original end of the River Eridanus, its name having been derived from the same roots as the similarly-named Achernar. Stars to the south of Acamar were inaccessible to Greek astronomers and it wasn't until explorers traveled south of the equator that the line of stars forming Eridanus was seen to extend into the southern sky. Consequently, new stars were added, extending the original Eridanus and resulting in a new termination point marked by Achernar.

Zaurak (γ Eridani) is a red giant star located near the northern end of Eridanus. The name of this star is appropriate, deriving as it does from the Arabic *'zauraq'* meaning 'Boat'. Shining at magnitude 2.97, the light from this star set off on its journey towards us around 200 years ago.

Rana (δ Eridani) is a magnitude 3.52 orange giant star lying at a distance of 29 light years and located immediately to the east of Epsilon Eridani. The name of this star means 'frog' in Latin, in keeping with its presence on the shores of the River Eridanus.

Epsilon (ε) Eridani is one of our closest stellar neighbors, the magnitude 3.72 glow from this orange dwarf star reaching us from a distance of just 10.5 light years.

Zibal (ζ Eridani) can be seen a little to the west of Epsilon Eridani and is a magnitude 4.80 white giant star whose light has taken 110 years to reach us.

Azha (η Eridani) is located slightly to the west of Zibal, the light from this magnitude 3.89 orange giant star having set off on its journey towards us around 135 years ago.

Angetenar ($τ^2$ Eridani) is a yellow giant star whose name is derived from the Arabic for **'the Bend in the River'** which is where it is seen to lie. The magnitude 4.76 glow of Angetenar reaches us from a distance of around 190 light years.

Moving along the River Eridanus we can see that there are several wide pairs of stars scattered along its length, including **Upsilon^1 (u^1)** and **Upsilon^2 (u^2) Eridani**. Shining at magnitude 3.81 from a distance of 214 light years, the yellow giant Upsilon^2 is marginally brighter than its magnitude 4.49 orange giant neighbor Upsilon^1, the light from which has taken around 125 years to reach us.

Immediately to the south of these is another pair comprising the magnitude 3.97 orange giant **Upsilon^3 (u^3)** and its magnitude 3.55 neighbor

Upsilon^4 (u^4) Eridani. Located at distances of 296 light years and 178 light years respectively, these two stars, along with nearby Upsilon^1 and Upsilon^2 form one of the several sweeping bends in the River Eridanus.

Beid (o^1 Eridani) is a magnitude 4.04 white giant star located at a distance of around 120 light years. The name of this star is derived from the Arabic *'al-baid'* meaning 'the Eggs'. Keeping with the egg-related theme is **Keid (o^2 Eridani)** which can be found immediately to the southeast of Beid. Deriving its name from the Arabic *'al-qaid'* meaning 'the Egg Shells', Keid shines at magnitude 4.43 from a distance of just 16.2 light years.

A mention of Keid brings us on to double stars, of which there are many to see in Eridanus. **Keid** is a good example, its magnitude 4.5 and 9.7 components being resolvable in small telescopes. The fainter companion is interesting in that it is itself a binary system comprised of white and red dwarf stars. The white dwarf component is one of the few stars of its type visible through small telescopes and is probably the easiest one to observe. What you are seeing are the small, hot remains of a collapsed star. White dwarfs are not actually producing new heat internally, but radiating it through the energy created by the collapse. The material forming a white dwarf is highly compressed by the forces of gravity and, as a result, stars of this type are incredibly dense. Generally containing as much mass as a star comparable in size to our own Sun, the material in a typical white dwarf is so compressed as to form an object not much larger than the Earth.

Amongst the other double stars found in Eridanus is **Acamar** *(also see left)*, which is widely regarded as being one of the finest double stars in the southern sky. The magnitude 3.4 and 4.4 blue-white components of this fine double are easily resolvable in a small telescope.

FORNAX
The Furnace

The whole of this tiny constellation is visible to stargazers located anywhere to the south of latitude 50°N but is not particularly prominent, so patience may be required to track it down.

The tiny and rather shapeless constellation Fornax takes the form of a short bent line of faint stars adjoining the western border of Eridanus, both constellations being depicted on the same chart *(see page 108)* and on the detailed chart here. It was devised by the French astronomer Nicolas Louis de Lacaille during his stay at the Cape of Good Hope in 1751/52, the purpose of his visit being to determine telescopically the positions of stars in the southern skies. While he was there he catalogued 9,800 individual stars and created a number of new constellations, all of which have found their way onto modern star charts. Lacaille introduced Fornax to depict a chemical furnace and to highlight the importance of chemistry to his

contemporaries although, as with many of the groups devised by Lacaille, the constellation bears little resemblance to the object it depicts.

BELOW: The superb barred spiral galaxy NGC 1365 in Fornax.

OPPOSITE: A Hubble Space Telescope image revealing prominent dust lanes and star clusters in the elliptical galaxy NGC 1316 in Fornax.

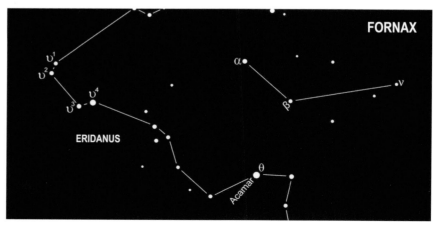

THE STARS OF FORNAX

Alpha (α) Fornacis is the brightest star in the constellation, shining at magnitude 3.80 from a distance of just 46 light years.

Beta (β) Fornacis is a magnitude 4.45 yellow giant star, the light from which has taken 173 years to reach us.

Nu (ν) Fornacis s shines at magnitude 4.68 from a distance of 370 light years.

GEMINI
The Twins

Located to the northeast of Orion are the two bright stars Castor and Pollux, the leading stars in the constellation of Gemini. Lying a little to the north of the celestial equator, the entire constellation can be viewed from virtually every inhabited part of the world. The two leading stars of Gemini are very distinctive and from here the rest of the group can be seen trailing away to the west of Castor and Pollux and in general direction of Orion.

Gemini is one of the oldest groups in the heavens and was depicted on the star charts of Babylonian astronomers as long ago as 1,500 BC. Greek mythology identifies the constellation as representing Castor and Pollux, the twin sons of Zeus and Leda, the two stars Castor and Pollux marking the heads of the twins. One of the twins, Castor, was a mortal and was slain in combat with Lynceus, the son of Aphareus. Pollux, who was immortal,

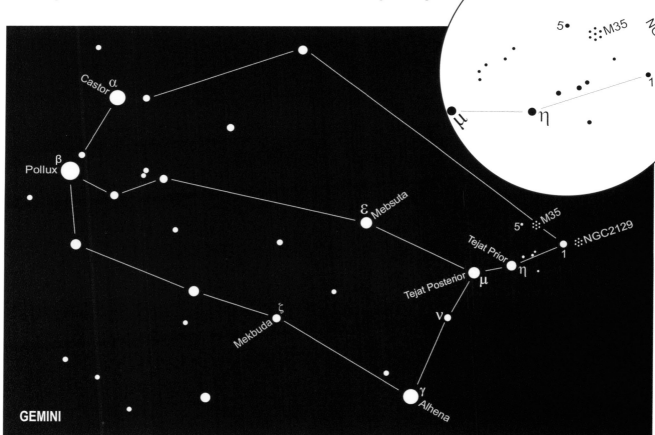

GEMINI

pleaded with Zeus that he be allowed to die so that he wouldn't be parted from his beloved brother, Zeus being so moved by this request that he placed both brothers in the sky so that they could be together for all time.

OPEN STAR CLUSTERS M35 AND NGC 2129

The three stars Nu Geminorum, Tejat Prior and Tejat Posterior are seen against the backdrop of the Milky Way which sweeps through the western boundary of Gemini, and checking out this region of sky with binoculars or a small telescope will reveal many star fields. One object of particular interest is the open star cluster **Messier 35** (M35) or NGC 2168, seen just to the northwest of Tejat Prior and

THE STARS OF GEMINI

Castor (α Geminorum) is the second brightest star in Gemini and shines with a magnitude of 1.58 from a distance of around 50 light years. Castor is a binary system with two companion stars in orbit around each other, each orbit taking over 450 years to complete. It was the Italian astronomer Giovanni Cassini who first detected that Castor was double in 1678, the two main components of the Castor system being fairly close to each other and requiring large telescopes to resolve the pair. In more recent times, a faint red dwarf star has been discovered to orbit the main pair, although what makes Castor even more unusual is that each of these three stars has since been found to be a binary in its own right, making Castor a rather elaborate sextuple star system!

The brightest star in Gemini is **Pollux (β Geminorum)** which shines at magnitude 1.16 from a distance of around 35 light years. Pollux has a true luminosity of over 40 times that of our own Sun, its yellow-orange tint easily seen in binoculars and contrasting with the whiteness of Castor.

Magnitude 1.93 **Alhena (γ Geminorum)** lies at a distance of just over 100 light years and has a true luminosity of around 160 times that of our Sun.

Mebsuta (ε Geminorum) shines with a magnitude of 3.06, the light from this yellow supergiant star having reached us from a distance of around 850 light years.

Another yellow supergiant is **Mekbuda (ζ Geminorum)** which is slightly fainter and more distant than Mebsuta, its magnitude 4.01 glow reaching us from a distance of over 1,200 light years.

Nu (ν) Geminorum is a blue giant star, the magnitude 4.13 glow of which has reached us from a distance of nearly 550 light years.

Tejat Prior (η Geminorum) is a red giant star shining at magnitude 3.31 from a distance of around 375 light years.

Located immediately to the west of Tejat Prior is **Tejat Posterior (μ Geminorum)**, another red giant whose ruddy glow, like that of Tejat Prior, can be detected in binoculars. Tejat Posterior shines at magnitude 2.87, its light having taken 230 years to reach us.

forming a triangle with Tejat Prior and the nearby star 1 Geminorum.

Visible with the unaided eye under exceptionally dark, clear skies, M35 appears as a faint, misty patch of light immediately to the northeast of the star 1 Geminorum. Containing well over 100

LEFT: Open star cluster M35 (NGC 2168) in Gemini.

stars and located at a distance of around 2,800 light years, M35 was discovered independently during the 1740s by the English astronomer John Bevis and the Swiss astronomer Jean-Philippe Loys de Cheseaux. Charles Messier added it to his catalogue in 1764, describing it as: 'A

OPPOSITE: The Medusa Nebula, a planetary nebula in Gemini, with the open star cluster NGC 2395 visible to the upper right of picture.

BELOW: Another open star cluster in Gemini is NGC 2158, seen here immediately to the upper right of M35.

cluster of very small stars near the left foot of Castor, a little distance from the stars Mu (Tejat Posterior) and Eta (Tejat Prior) of that constellation . . . '. Binoculars will bring out M35 quite well and may reveal some of the individual stars making up the cluster.

Powerful binoculars, or a small telescope, may also reveal another open cluster, this being **NGC 2129**, located more or less on the opposite side of the star 1 Geminorum as shown. Shining from a distance of over 7,000 light years. NGC 2129 may be a little difficult to spot, although with perseverance you should pick it up. This finder chart *(see page 112)*

shows the area of sky around these two star clusters in more detail, the individual stars shown being those visible through binoculars. Use the finder chart to help you star hop your way from Tejat Posterior (μ) through Tejat Prior (η) and on towards your target. As is the case with M35, when viewed through binoculars NGC 2129 will appear as a faint, misty patch of light, although a small telescope may resolve some of the individual stars within the cluster.

GRUS
The Crane

The constellation Grus lies immediately to the south of Piscis Austrinus (the Southern Fish). The bright star Fomalhaut in Piscis Austrinus is included here as a guide to locating Grus, the whole of which is visible from latitudes south of 33°N.

This is of the groups introduced by the Dutch navigators and explorers Pieter Dirkszoon Keyser and Frederick de Houtman following their expedition to the East Indies in the 1590s. It depicts the long-necked bird, a crane, and first appeared on the celestial globe produced by Petrus Plancius in 1598 as well as in the star atlas *Uranometria* produced by Johann Bayer in 1603. Grus was formed from stars located immediately to the south of Piscis Austrinus and which originally formed the tail of the Southern Fish.

GRUS

THE STARS OF GRUS

Bluish-white **Alnair (α Gruis)** is the brightest star in Grus, shining with a magnitude of 1.73 from a distance of 101 light years. Alnair derives its name from an abbreviation of the Arabic for 'the Bright One from the Fish's Tail' a reminder that the stars in this region once formed the tail of Piscis Austrinus.

In keeping with this theme, the name of the magnitude 3.00 **Al Dhanab (γ Gruis)** is derived from the Arabic for 'the Tail'. The light from Al Dhanab has taken just over 210 years to reach us.

The magnitude 2.07 red giant star **Beta (β) Gruis** shines from a distance of around 175 light years.

TWO NAKED-EYE DOUBLE STARS

The two yellow giant stars **Mu[1] (μ[1])** and **Mu[2] (μ[2])** form a naked-eye double. Although these two stars both lie roughly 270 light years away, observation has shown that they are not gravitationally linked and simply happen to lie in more or less the same line of sight as seen from our planet.

Located a little way to the southeast of the Mu pair are the two stars **Delta[1] (δ[1])** and **Delta[2] (δ[2])** which also form a wide naked eye optical double. The magnitude 3.97 yellow giant Delta[1] shines from a distance of around 310 light years, putting it a little closer than the magnitude 4.12 red giant Delta[2], the light from which has taken around 330 years to reach us.

HERCULES
Hercules

Hercules covers such a large area of sky that portions of the constellation can be seen from almost every inhabited part of the world, the entire group being visible to observers anywhere north of latitude 39°S. It contains no really bright stars, although the quadrilateral of four stars **Epsilon (ε), Pi (π), Eta (η)** and **Zeta (ζ)** marking the central region of Hercules is quite distinctive. These four stars are collectively known as the 'Keystone' and, once identified, the rest of the stars of Hercules can be traced out around it

The brilliant star Vega in the constellation Lyra lies just to the east of Hercules while the circlet of stars forming Corona Borealis, together with its leading star Alphecca, lies immediately to the west. Taking our gaze further west we arrive at the brilliant star Arcturus in Boötes. A line taken from Arcturus, passing through Alphecca in Corona Borealis and on as far again will lead you to the 'Keystone'. The three stars Arcturus, Alphecca and Vega are included on the chart to help you locate the Keystone and, from there, the rest of the sprawling form of Hercules.

The Greeks knew this constellation as *Engonasin*, meaning 'The Kneeling

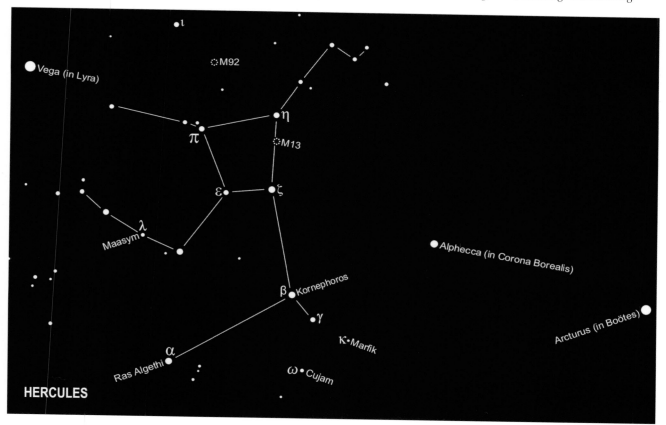

THE STARS OF HERCULES

The brightest star in Hercules is the magnitude 2.78 yellow giant **Kornephoros (β Herculis)**, the light from which has taken around 140 years to reach us. Kornephoros derives its name from the Greek for 'the Club Bearer', a name that was once applied to the constellation as a whole. The red supergiant **Ras Algethi (α Herculis)** is located a little to the southeast of Kornephoros. Shining from a distance of around 350 light years and taking its name from the Arabic *'ra's al-jathi'* meaning 'the Kneeler's Head', Ras Algethi is an irregular variable star, the light from which varies between 3rd and 4th magnitudes. Ras Algethi has a distinct orange-red tint which should be visible in binoculars.

Magnitude 2.81 **Zeta (ζ) Herculis** lies at a distance of 35 light years, less than a tenth of the distance of **Pi (π) Herculis**, a magnitude 3.16 orange giant whose light has taken around 380 years to reach us. The magnitude 3.92 glow of **Epsilon (ε) Herculis** reaches us from a distance of just over 150 light years while that of magnitude 3.48 yellow giant **Eta (η)**

Herculis set off towards us around 110 years ago.

Maasym (λ Herculis) is an orange giant star, the magnitude 4.41 glow from which reaches us from a distance of 369 light years.

Deriving its name from the Latin for 'club', **Cujam (ω Herculis)** shines at magnitude 4.57 from a distance of around 250 light years.

Marfik (κ Herculis) takes its name from the Arabic *'al-marfiq'* meaning 'the Elbow' and is a double star located a little way to the southwest of Gamma (γ) Herculis. The brighter component shines at magnitude 5.3 from a distance of around 365 light years, the yellowish tint of this star contrasting with the bronze hue of the magnitude 6.5 companion. The proximity of these two stars to each other is merely a line of sight effect, the light from the fainter star having taken over 450 years to reach us. Both components may just be resolved in good binoculars although a telescope may be needed in order to detect their individual colors.

THE GREAT HERCULES CLUSTER

Globular clusters are huge spherical collections of stars which lie in the region of space around our Galaxy *(see Glossary)*, the Keystone of Hercules being home to one of the finest globular clusters in the northern sky. Also known as the Great Hercules Cluster, **Messier 13** (M13) or NGC 6205 measures around 150 light years in diameter and is thought to contain anything up to half a million stars. This celestial showpiece was discovered by the English astronomer Edmund Halley in 1714 and catalogued by Charles Messier in 1764. The true nature of this object was not evident to Halley, who described it as ' *. . . a little patch . . .* ', Messier seeing only *'A nebula which I am sure contains no star.'* It was left to the large telescopes of William Herschel to reveal the true nature of this object, his view of M13 prompting Herschel to describe it as *'A most beautiful cluster of stars . . .'.*

One' and it wasn't until the third century BC that the Greek astronomer Eratosthenes of Cyrene identified it as representing Hercules, the legendary character who carried out 12 labours at the command of Eurystheus, King of Argos. Hercules was possessed of great strength and courage and was known far and wide for his exploits, a reputation that was enhanced by the convincing way in which he carried out the dangerous tasks given to him by Eurystheus, which included destroying the fearsome multi-headed Lernaean Hydra and slaying the dragon Ladon.

With an overall magnitude of 5.8, M13 shines from a distance of around 25,000 light years and may be visible as a faint, misty patch of light to the naked eye, providing the sky is dark, clear and free of moonlight. Located roughly a third of the way on a line between Eta and Zeta, binoculars or a small telescope will easily pick out the faint glow from M13, revealing it as a fuzzy, circular patch of light. Larger telescopes would be needed in order to resolve individual stars within the cluster.

GLOBULAR CLUSTER M92

The presence of the Great Hercules Cluster often draws attention away from another globular located in Hercules. This is **Messier 92** (M92) or NGC 6341 which lies below naked-eye visibility, its magnitude 6.3 glow requiring the use of an optical aid to locate it.

M92 can be found roughly two-thirds of the way along a line taken from Eta towards Iota (ι) Herculis. Containing around 250,000 stars and lying at a distance of just under 27,000 light years, this cluster has a diameter of around 110 light years and is visible in binoculars or a small telescope as a non-stellar, fuzzy patch of light. M92 was discovered in 1777 by the German astronomer Johann Elert Bode and independently rediscovered by

OPPOSITE: Globular cluster M92 (NGC 6341) in Hercules.

ABOVE RIGHT: The crowded central regions of the Great Hercules Cluster M13 (NGC 6205).

Charles Messier in 1781 who subsequently added it to his catalogue. However, as was the case with M13, it was William Herschel who first resolved individual stars within the cluster.

It is a sobering thought that, when you look at M92, or its somewhat grander neighbor M13, the light you are seeing from these remote objects set off towards us around five times as long ago as the period during which construction began on Stonehenge and the Great Pyramid of Giza; the universe is truly a huge place!

HOROLOGIUM
The Pendulum Clock

Visible in its entirety from latitudes south of 23°N, Horologium is another of the constellations introduced by the French astronomer Nicolas Louis de Lacaille. It comprises a meandering line of faint stars lying immediately to the east of the bright star Achernar in the bordering constellation Eridanus and which is included on the chart for guidance. As with most of the other Lacaille-generated constellations, Horologium looks nothing like the object it is supposed to depict.

The globular cluster NGC 1261 in Horologium was discovered by the Scottish astronomer James Dunlop in 1826.

THE STARS OF HOROLOGIUM

The magnitude 3.85 orange giant **Alpha (α) Horologii** is located at a distance of 115 light years and forms a wide pair with magnitude 4.93 **Delta (δ) Horologii**. These two stars, which lie at the northeastern extremity of Horologium, are not physically connected, the light from Delta having taken around 180 years to reach us.

Beta (β) Horologii can be found at the opposite end of the constellation, the feeble magnitude 4.98 glow of this white giant star having reached us from a distance of around 300 light years.

Located a little to the north of β is magnitude 5.12 **Mu (μ) Horologii**, the light from which star set off on its journey towards us around 140 years ago.

The trio of 5th magnitude stars **Zeta (ζ) Horologii, Eta (η) Horologii** and **Iota (ι) Horologii Horologii** make up the central portion of this uninspiring constellation.

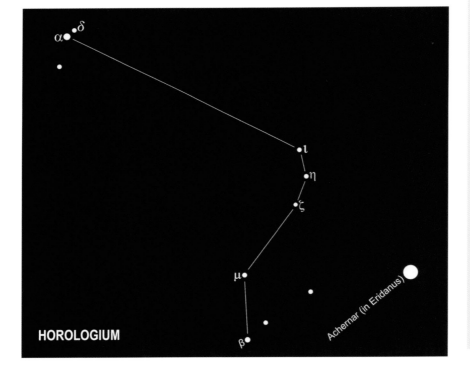

HOROLOGIUM

HYDRA
The Water Snake

The Head of the long and sprawling constellation Hydra lies in the region of sky to the southwest of the bright star Regulus (in Leo) and immediately to the north of the celestial equator. From here the snake winds its way eastwards, eventually reaching the area to the south and southeast of the bright star Spica (in Virgo). (The stars Regulus and Spica are depicted on the chart as guides to help you locate and follow the constellation along its whole length.) Hydra can be seen in its entirety from anywhere south of latitude 55°N, with portions of the constellation being visible from anywhere. The constellation is best seen during the months of March and April when its winding form can be seen straddling an extended region of sky on and to the south of the celestial equator.

Hydra is one of the 48 star groups drawn up by the Greek astronomer Ptolemy during the second century and is the largest of the 88 modern-day constellations. The stars which form Hydra, however, are not particularly bright and backyard astronomers observing from mid-northern latitudes will need clear and dark skies in order to

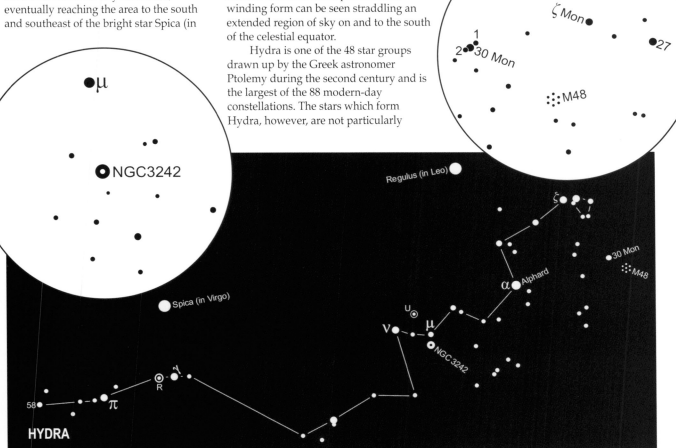

THE STARS OF HYDRA

Alphard (α Hydrae) is the brightest star in Hydra. Deriving its name from the Arabic **'al-fard'** meaning 'the Solitary One', Alphard is an orange giant shining at magnitude 1.99 from a distance of 180 light years.

Situated at the eastern end of the Head of Hydra is **Zeta (ζ) Hydrae**, a magnitude 3.11 yellow giant star located at a distance of over 160 light years.

Proceeding along the constellation we soon come to the two orange giant stars **Nu (ν)** and **Mu (μ) Hydrae** which together act as a guide to the variable star U Hydrae (see opposite). Of these two stars Nu is the closest, shining at magnitude 3.11 from a distance of 144 light years, the light from magnitude 3.83 Mu having set off on its journey towards us a little over 230 years ago.

Gamma (γ) Hydrae lies a little way to the south of the bright star Spica in the neighboring constellation Virgo. Magnitude 2.99 Gamma is a yellow giant whose light has taken over 130 years to reach us. Spica will be useful in helping you locate Gamma as well as the variable star R Hydrae (see opposite).

The winding form of Hydra continues eastwards via the magnitude 3.25 orange giant **Pi (π) Hydrae** and eventually ends with magnitude 4.42 **58 Hydrae** which, located at a distance of around 330 light years, depicts the tail of the celestial water snake.

identify the entire group. Once you have managed to locate the Head of Hydra you can trace out the rest of the constellation with binoculars, working eastwards and picking out its main stars. Those at more southerly latitudes should have less trouble as Hydra will be situated fairly high in the northern sky and away from horizon glow.

According to one account, Hydra represents the fearsome multi-headed monster which lived near to the town of Lerna in the Pelopponesus region of southern Greece and which Hercules killed as the second of his labors. According to another legend, Apollo asked a crow to fly to a spring and fetch him the water of life, giving him a cup in which to carry the precious cargo. While en route, the crow stopped to eat some fruit from a fig tree, during which process he dropped the cup onto the ground. The crow panicked and, retrieving the cup took it, together with a water snake, back to Apollo. The snake was blamed for the delay in extracting water from the spring, although Apollo did not believe a word of

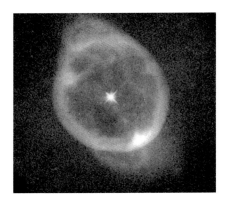

the crow's excuses. Annoyed at what had happened, Apollo banished the crow, cup and water snake to the sky where both the crow and cup *(see Corvus and Crater)* are forever balanced on the back of the water snake.

PLANETARY NEBULA NGC 3242 – THE GHOST OF JUPITER

Discovered by William Herschel in 1785, the planetary nebula **NGC 3242** lies a little to the south of the star Mu (μ) Hydrae. Shining with a magnitude of around 8.5 from a distance of around 1,500 light years, NGC 3242 can be tracked down with the finder chart *(see page 121)* which shows Mu Hydrae along with a number of fainter stars in the immediate area of the nebula. Look for this object either with a good pair of binoculars or a small telescope, remembering to look for a diffuse patch of light rather than a star-like point.

Planetary nebulae are created when gas is ejected from old stars. They take the appearance of shells of gas surrounding the parent star, the gas within them illuminated by energy from the star from whose outer layers they were formed. The term 'planetary nebula' was devised by William Herschel who likened their appearance in telescopes to that of planetary discs. Planetary nebulae occur in a variety of colors and shapes, NGC 3242 itself often being referred to as the 'Ghost of Jupiter', presumably due to the fact that early observers likened its visual appearance to that of the planet Jupiter.

LEFT: Planetary nebula NGC 3242 in Hydra.

OPPOSITE: Open star cluster M48 (NGC 2548) in Hydra.

Monocerotis which were mentioned by Messier and which are also shown on the star chart for Monoceros *(see Monoceros)*. Using the finder chart you can star hop your way to M48 which will at first appear as an extended patch of faint light, although closer and more careful examination will bring out some of its member stars.

TWO VARIABLE STARS IN HYDRA

Hydra plays host to a couple of interesting variable stars, one of which is the semi-regular **U Hydrae** which can be located fairly easily due to it forming a small but distinct triangle with nearby Nu and Mu. U Hydrae has the noticeably reddish tint typical of stars of this type and a period of around 115 days, during which time it ranges in brightness between around 4th and 5th magnitudes.

The Mira-type variable **R Hydrae** is one of the easiest variable stars to observe. It can be found just to the east of Gamma, within the same binocular field of view, and is regarded as one of the easiest variable stars to observe. The variability of R Hydrae is generally thought to have been discovered by the Italian-born astronomer Giacomo Filippo Maraldi in 1704, and estimates as to how much this star varies in brightness differ. Its average magnitude range is between around 4.5 to 9.5 with the whole cycle taking place over a period of around 385 days. R Hydrae is easy to find and is an ideal candidate for backyard astronomers. Binoculars or a small telescope will allow you to observe its complete cycle of variability, and you may even spot R Hydrae with the naked eye when at or near its maximum brightness.

OPEN STAR CLUSTER M48

To the southwest of the Head of Hydra is the star 30 Mon, slightly beyond which is the open star cluster **Messier 48** (M48) or NGC 2548. Containing up to 80 member stars and shining with an overall magnitude of around 5.5, this cluster is just about visible to the naked eye under exceptionally dark and clear skies.

Originally discovered in 1771 by the French astronomer Charles Messier, the cluster was described by him as being '. . . *a little distance from the three stars which are at the root of the tail of Monoceros'*.

To locate M48 in binoculars use the star 30 Mon as a guide. The finder chart *(see page 121)* depicts 30 Mon along with the three stars 27, 28 and Zeta (ζ)

HYDRUS
The Little Water Snake
(See Tucana)

INDUS
The Indian

Indus lies immediately to the southwest of the constellation Grus and is visible in its entirety from latitudes south of 15°N. The constellation represents a Native American and is another of the groups introduced by the Dutch navigators and explorers Pieter Dirkszoon Keyser and Frederick de Houtman following their expedition to the East Indies in the 1590s.

None of the stars in Indus are particularly bright and none are named. The star Alnair in Grus is shown here as a guide to locating this faint constellation, along with the star Peacock in the constellation Pavo, which can be found to the southwest of Indus.

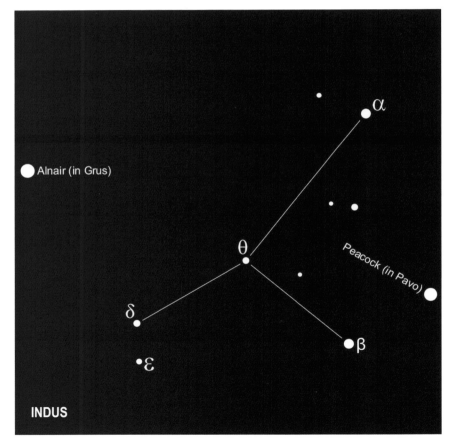

Alnair (in Grus)

Peacock (in Pavo)

α

θ

δ

ε

β

INDUS

OPPOSITE: A Hubble Space Telescope image of NGC 7049, a galaxy in Indus. The unusual appearance of this galaxy is largely due to the prominent dust ring seen superimposed against the starlight behind it. The bright star visible at the top of the ring is actually a foreground star located in our own Galaxy.

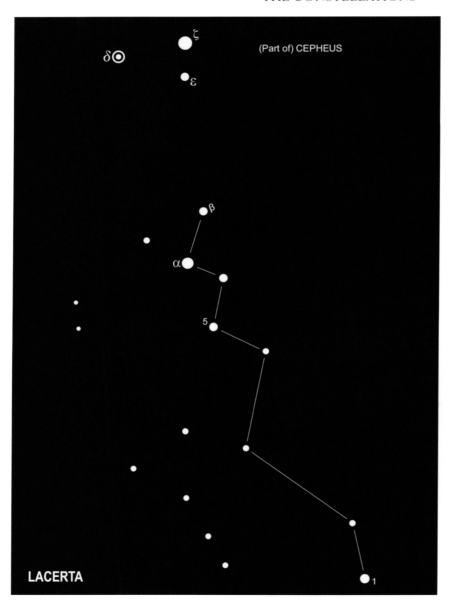

ζ

δ ◉

ε

(Part of) CEPHEUS

β

α

5

LACERTA

1

THE STARS OF LACERTA

Alpha (α) Lacertae is the brightest of the stars forming this constellation, its feeble magnitude 3.76 glow reaching us from a distance of around 100 light years.

To the northwest of Alpha we find the yellow giant **Beta (β) Lacertae**, a magnitude 4.42 star located at a distance of 170 light years.

The magnitude 4.34 orange supergiant star **5 Lacertae** shines from a distance in excess of 1,000 light years.

1 Lacertae, a magnitude 4.14 orange giant whose light has taken just over 600 years to reach us, marks the southern extremity of Lacerta.

ABOVE:Open star cluster NGC 7243 in Lacerta.

OPPOSITE: An extensive region of gas and dust located on the border of Lacerta and the adjoining constellation Pegasus.

LACERTA
The Lizard

Lacerta was introduced in the 17th century by the Polish astronomer Johannes Hevelius in order to fill a gap in the barren region to the north of Pegasus. The whole of the constellation can be seen from locations north of latitude 33°S although, because there are no particularly bright stars within this group, observers south of the equator may have difficulty seeing it at all unless the sky above their northern horizon is dark, clear and free of light pollution.

The constellation takes the form of a zigzag line of stars running from north to south situated between Pegasus to the south, Cygnus to the west, Cassiopeia to the northeast and Cepheus immediately to the north. The triangle formed from the three stars Delta (δ), Zeta (ζ) and Epsilon (ε) in Cepheus is shown here to help you locate this faint group.

None of the stars in Lacerta have any names and there are no legends attached to it. The constellation is very faint with no stars brighter than 4th magnitude although, providing the sky is dark and clear, and you arm yourself with a pair of binoculars, you should have no problem locating the Lizard's wandering form.

127

LEO
The Lion

When seen from mid-northern latitudes, the conspicuous form of Leo occupies a position high up in the southern sky during March and April and is one of the principal constellations of northern spring. Equally prominent for southern hemisphere observers, Leo graces the northern heavens and, situated as it is a little way to the north of the celestial equator, can be seen in its entirety from almost every inhabited part of the world.

Of the numerous accounts as to how the lion came to be in the sky, perhaps the best known is the Greek legend which identifies Leo as the Nemean Lion which Hercules slew as the first of his 12 labors. Leo is one of the few constellations that actually resembles the object or character that it depicts, the group indeed echoing the appearance of a crouching lion. Its most recognizable feature is the curve of stars extending northwards from Regulus. Often known as the Sickle, this curve of stars depicts the head, mane and paws of the lion, the triangle of stars formed from Denebola, Coxa and Zosma representing the lion's hindquarters and tail.

PAGE 130: Zooming in on the magnificent spiral galaxy M66 (NGC 3627) in Leo.

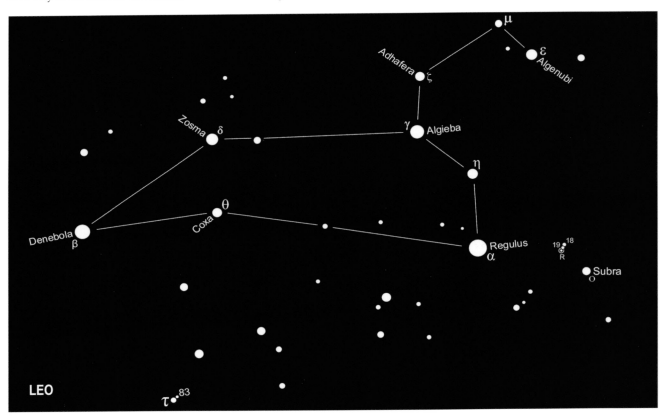

THE STARS OF LEO

Leo contains over 100 naked-eye stars, the brightest of which is **Regulus (α Leonis)** which takes its name from the Latin for 'the Little King', a title which was first applied to the star during the early 16th century. Shining from a distance of around 80 light years, magnitude 1.36 blue-white Regulus has a fainter 8th magnitude yellowish companion lying quite close by, both stars being resolvable in small telescopes.

Regulus has long been held as a royal star, not least by the astronomer Joseph Henry Elgie who wrote of it that: *'The nomenclature of many stars has often seemed to me hauntingly melodious, no matter how bizarre the names should appear in cold type. I would instance Antares, Arcturus, Betelgeuse, Vega, Capella, Canopus. Then how resonantly royal is the name of Regulus!'*

Deriving its name from the Arabic **'al dhanab al-asad'** meaning 'the Lion's Tail', **Denebola (β Leonis)** is a white magnitude 2.14 star located 36 light years away. Denebola, along with the nearby bright stars Arcturus in Boötes, Spica in Virgo and Cor Caroli in Canes Venatici, form the conspicuous and celebrated asterism the Diamond of Virgo, which surrounds the constellation Coma Berenices *(see Coma Berenices)*.

The hindquarters of Leo are formed from Denebola and the two stars **Coxa (θ Leonis)** and **Zosma (δ Leonis)**. Magnitude 2.56 Zosma shines from a distance of 58 light years, brighter and somewhat nearer than its magnitude 3.33 neighbor Coxa, the light from which set off towards us over 160 years ago.

Eta (η) Leonis is a white supergiant star shining at magnitude 3.48 from a distance in excess of 1,200 light years.

The magnitude 3.52 glow of **Subra (o Leonis)** reaches us from a distance of 130 light years.

The orange-red **Algieba (γ Leonis)** derives its name from the Arabic **'al jabha'** meaning 'the Lion's Forehead or Mane'. Algieba has a magnitude of 2.3 and its orange tint offers a contrast with a fainter magnitude 3.5 yellowish companion star. Both Algieba and its companion can just be resolved in a small telescope, although they may present something of a challenge!

The double star **Tau (τ) Leonis**, lying at a distance of around 600 light years and found a little way to the south of the star Coxa, is a target for the moderately-equipped backyard astronomer, comprising as it does a brighter yellow-orange star with a fainter bluish companion, both of which should be visible in binoculars. Lying to the immediate northwest of Tau is the star **83 Leonis**, which a telescope will reveal is also a double, its yellowish and reddish components shining from a distance of almost 60 light years, roughly a tenth the distance of Tau.

Algenubi (ε Leonis) shines at magnitude 2.97 from a distance of around 250 light years and derives its name from the Arabic **'ras al-asad al-janubi'**, meaning 'the Southern (Star) in the Lion's Head'. Algenubi, along with nearby **Mu (μ) Leonis** and **Adhafera (ζ Leonis)**, forms the main outline of the lion's mane. Mu and Algenubi are also respectively known as **Ras Elased Borealis** and **Ras Elased Australis**, names of Arabic origin which allude to the positions of these two stars in the lion's head.

Leo is also home to the Mira-type *(see entry for Cetus)* variable star **R Leonis**, one of the most widely-observed objects of its type in the sky. Shining from a distance of around 380 light years, and located a little way to the west of Regulus, as shown on the chart, the variability of this star was first noted by the Polish astronomer Julius August Koch in 1782. R Leonis has an average magnitude range of between 5 and 10, the complete cycle of variability taking place over a period of 312 days. To find R Leonis, follow a line from Regulus to Subra, the pair of faint stars 18 and 19, along with R Leonis, being seen to lie just above this line. The characteristic ruddy hue of R Leonis should be evident when you see this star, which was described by the English astronomer Edwin Dunkin as having a *'. . . blood-red appearance, which is very striking to the eye when viewed for the first time through a good telescope'*.

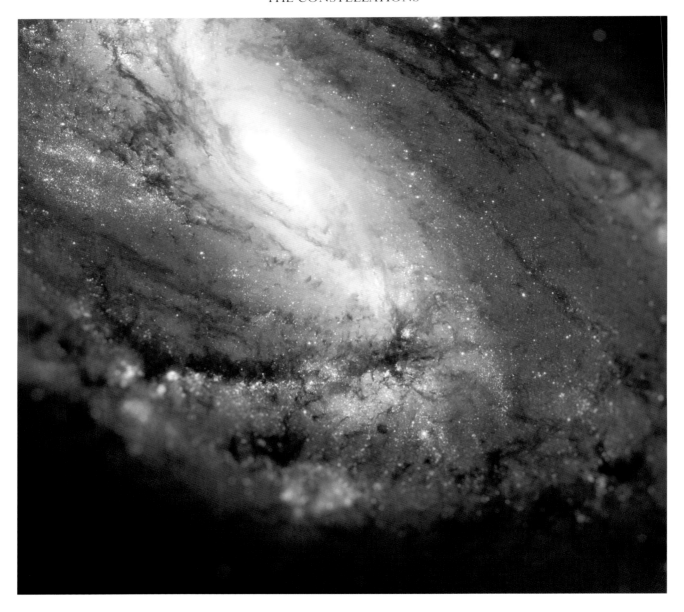

LEO MINOR
The Little Lion

Leo Minor takes the form of a short zig-zag line of stars running between Ursa Major and Leo and is another of the obscure constellations introduced during the 17th century by the Polish astronomer Johannes Hevelius. The whole of this constellation can be seen from locations to the north of latitude 48°S, ruling out only the southernmost regions of South America and New Zealand as potential vantage points.

The stars of Leo, some of which are shown here, can be used as a guide to locating Leo Minor. As is the case with many of the fainter constellations, Leo Minor can be seen with the naked eye, although tracking it down would be made easier with the help of binoculars. Because there are no bright stars in the constellation, observers south of the equator may have difficulty picking it out unless their northern sky is dark and clear and free of any form of light pollution.

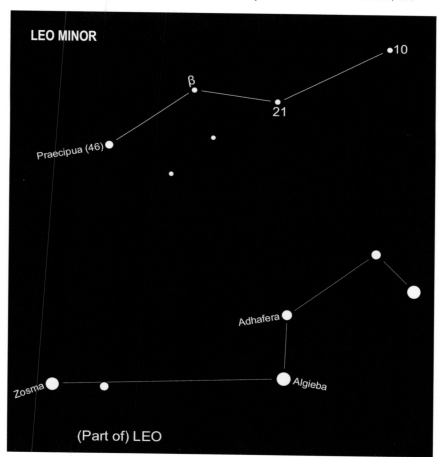

THE STARS OF LEO MINOR

The brightest star in Leo Minor is **Praecipua**, a name which means 'Chief' and which was given to the star by Hevelius. Located at the extreme eastern end of Leo Minor, Praecipua is an orange giant shining with a magnitude of 3.79, its light having taken 95 years to reach us. The name Praecipua never became popular, and this star is usually known by its Flamsteed number 46.

Beta (β) Leonis Minoris is a yellow giant star lying at a distance of just over 150 light years and shining at magnitude 4.20.

10 Leonis Minoris and **21 Leonis Minoris** complete the constellation. Shining from a distance of around 185 light years, and lying at the western end of Leo Minor, the yellow giant star 10 glows at magnitude 4.54, slightly dimmer than 21 which, at magnitude 4.49, is located at a distance of 90 light years.

LEPUS
The Hare

Lepus can be viewed in its entirety from latitudes south of 63°N, which puts it within the reach of observers in central Canada, northern Europe and central Russia and from anywhere further south of these regions. The distinctive shape of Lepus is easy to track down, being located immediately to the south of Orion, the two stars Rigel and Saiph in Orion being included on the chart to help in locating the group.

Accounts as to the origins of Lepus vary, Arabic astronomers having likened the group to a herd of thirst-slaking camels who were drinking from the nearby Milky Way. The Egyptians identified the stars in Lepus as the legendary Boat of Osiris, the powerful Egyptian god whose form was represented by the nearby Orion. Another story comes from the Greeks of Sicily and tells of the great devastation to crops wrought by the local hare population. To remedy this they placed the Hare up in the sky close to the mighty hunter Orion, in the expectation that he could keep their numbers under control.

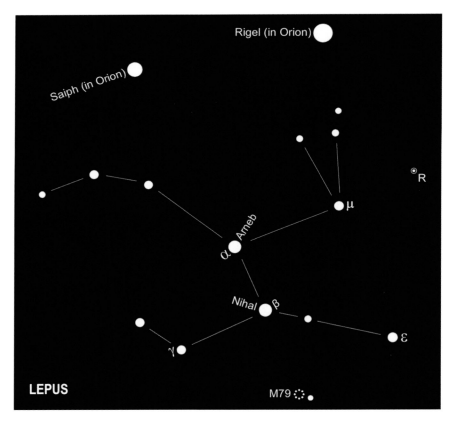

THE STARS OF LEPUS

Arneb (α Leporis) is the brightest star in Lepus, this white supergiant shining at magnitude 2.58 from a distance of a little over 2,000 light years. Arneb derives its name from the Arabic **'al-arnab'** meaning 'the Hare'.

Nihal (β Leporis) can be found just to the south of Arneb, the magnitude 2.81 glow from this yellow giant star having set off on its journey towards us around 160 years ago.

The orange giant star **Epsilon (ε) Leporis** shines at magnitude 3.19 from a distance of a little over 200 light years.

Slightly fainter than Epsilon is **Mu (μ) Leporis**, the light from this magnitude 3.29 star reaching us from a distance of 186 light years

Gamma (γ) Leporis is a yellowish-white star shining at magnitude 3.59 from a distance of only 29 light years. Close examination with powerful binoculars or a small telescope will reveal that Gamma has a companion star with a distinctly orange tint and, when seen together, the pair form a pretty color contrast.

HIND'S CRIMSON STAR

The well-known variable star **R Leporis**, also known as Hind's Crimson Star, can be found near the western boundary of Lepus as shown. This is a Mira-type variable and is named after the prototype long-period variable Mira in the constellation Cetus, under which entry this type of star is described in more detail. R Leporis varies in brightness between magnitudes 5.5 and 11.5 over a period of around 430 days and lies at a distance of over 1,000 light years. Its variability was discovered by the English astronomer John Russell Hind in 1845 after whom the star gained its popular name. If you look for the star and have difficulty in finding it, this may well be because it is at or near minimum magnitude. If you keep checking, R Leporis will eventually come into view as it brightens again and will be distinguishable by its strong red color. You should be able to follow the star through its entire period of variability with either binoculars or a small telescope. However, it is worth remembering that variable stars of this type can take a year or more to undergo their complete cycles, so any observations you start now may not be complete until the following year!

GLOBULAR CLUSTER M79

Lepus plays host to the globular cluster **Messier 79** (M79) or NGC 1904 which can be found by extending a line from Arneb, through Nihal and approximately southwards for roughly the same distance again. M79 was discovered by the French astronomer Pierre François André Méchain in 1780 and was first resolved into stars and recognized as a globular cluster by William Herschel in around 1784. Herschel described it as ' . . . *a beautiful cluster . . . globular and extremely rich'*.

Lying at a distance of a little over 40,000 light years, M79 can be seen in binoculars or a small telescope through which it will appear as a hazy star-like object, although larger telescopes should reveal a number of its member stars. Moonlight or horizon mist or glow may tend to hide M79 from the view of observers at mid-northern latitudes, although for those further south the cluster will be higher in the sky and tracked down fairly easily.

BELOW: Globular cluster M79 (NGC 1904) in Lepus.

LIBRA
The Scales

Bordered by Virgo (the Virgin) to the northwest and Scorpius (the Scorpion) to the southeast is the constellation Libra, parts of which are visible worldwide, the whole of the constellation being observable from anywhere south of latitude 60°N. Although Libra is slightly larger than Scorpius, it contains no stars

brighter than 2nd magnitude and is by no means as prominent. Observers at mid-northern latitudes may have difficulty picking this group out at all unless the sky is fairly dark and clear. However, a careful search with binoculars of the region to the southeast of the neighboring constellation Virgo should help you track down the quadrilateral of stars that form the main section of Libra.

The stars in Libra were once known as Chelae, meaning 'The Claws (of the Scorpion)' and originally represented the claws of the neighboring Scorpius, this being reflected in the names of some of the stars in Libra. The early Greeks did not identify the constellation with Scales or a Balance, this association having been introduced by the Romans during the first century BC.

ZUBEN ELAKRIBI - AN ALGOL-TYPE VARIABLE

Located a little way to the west of Zubeneschamali and shining from a distance of almost 300 light years, **Delta (δ) Librae,** also known as **Zuben Elakribi,** is an Algol-type eclipsing binary (*see Perseus*) with a magnitude range of between 4.8 and 5.9 and a total period of just under 2 days and 8 hours. As with all variables of this type, this star is worth keeping an eye on in case you manage to detect a dip in its overall magnitude.

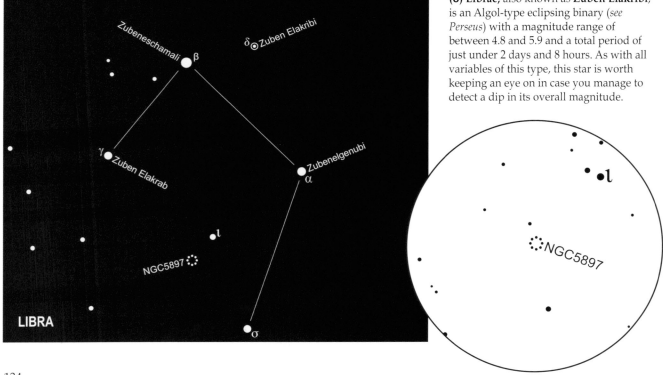

THE STARS OF LIBRA

Magnitude 2.61 **Zubeneschamali (β Librae)** is the brightest star in Libra. Deriving its name from the Arabic for 'The Northern Claw', the light from Zubeneschamali has taken 185 years to reach us.

Located at a distance of 76 light years and taking its name from the Arabic for 'The Southern Claw', **Zubenelgenubi (α Librae)** is a double star, its blue-white magnitude 2.7 primary and yellowish magnitude 5.2 secondary both being resolvable in binoculars.

Zuben Elakrab (γ Librae) is a magnitude 3.91 yellow giant star shining from a distance of around 160 light years.

The light from the magnitude 3.25 red giant star **Sigma (σ) Librae** reaches us from a distance of around 290 light years.

Iota (ι) Librae shines at magnitude 4.54 from a distance of 380 light years.

GLOBULAR CLUSTER NGC 5897

Located on a line between Zuben Elakrab and Sigma, a little to the southeast of Iota, is the globular cluster **NGC 5897**. Shining with an overall magnitude of 8.5 from a distance of around 40,000 light years, this cluster is rather faint and difficult to see but can be detected in good binoculars under really dark, clear skies. The finder

RIGHT: Globular cluster NGC 5897 in Libra.

chart shows the star field around Iota and NGC 5897. Once you have identified Iota, carefully search the area of sky immediately to its southeast (remembering to look for a patch of light rather than a star-like point) and you should be able to pick it up.

LUPUS

LUPUS
The Wolf

The area of sky to the east of Centaurus (the Centaur) is occupied by the constellation Lupus, the bright pair of stars Alpha (α) and Beta (β) Centauri depicted on the chart as guide stars to help you locate the group. Lupus was one of the 48 constellations drawn up by the Greek astronomer Ptolemy during the second century although there are no legends attached to it. The stars forming

THE STARS OF LUPUS

Located near the western borders of Lupus is **Alpha (α) Lupi**, the brightest member of the constellation. Shining at magnitude 2.30, the light from this star reaches us from a distance of a little over 450 light years.

Beta (β) Lupi shines at magnitude 2.68 from a distance of 382 light years.

The light from magnitude 2.80 **Gamma (γ) Lupi** set off on its journey towards us a little over 400 years ago.

Delta (δ) Lupi shines at magnitude 3.22 from a distance of around 860 light years.

The magnitude 3.37 glow of **Epsilon (ε) Lupi** reaches us from around 500 light years away.

Zeta (ζ) Lupi is a magnitude 3.41 yellow giant located at a distance of 117 light years.

closer than bluish Psi² which shines at magnitude 4.75 from a distance of around 360 light years.

Both the above pairs are particularly attractive when seen through binoculars, although a small telescope is required to resolve the two components of the binary **Eta (η) Lupi**. Shining from a distance of around 440 light years, the blue-white magnitude 3.6 primary is accompanied by a much fainter magnitude 7.8 secondary which has been described as having an ashy tint. Another excellent target for small telescopes is the double star **Xi (ξ) Lupi** which has magnitude 5.2 and 5.5 yellowish components.

OPEN STAR CLUSTER NGC 5822

Located a little to the south of Zeta (ζ) Lupi we find the open star cluster NGC 5822 which is an easy binocular object providing the sky is reasonably dark and clear. NGC 5822 shines with an overall magnitude of around 7 and contains around 100 stars ranging in magnitude from around 9 to around 12. Using binoculars, follow a line from Zeta southwards towards the bright star Alpha Centauri until you reach the cluster. Good binoculars or a small telescope will reveal some of the brightest members of NGC 5822.

this group were formerly combined with Centaurus and represented an animal impaled on a long pole held by the Centaur who was facing in the general direction of Ara (the Altar), presumably so that he could offer the animal to the gods as a sacrifice.

Lupus can be viewed in its entirety from the southern United States, Egypt, central China and from latitudes further south. None of the stars in Lupus is named and none is particularly prominent.

DOUBLE STARS IN LUPUS

Lupus contains a couple of double stars which are resolvable with the naked eye, one of which is the optical double formed from magnitude 3.57 **Phi¹ (φ¹)** and magnitude 4.54 **Phi² (φ²)**. The orange tint of **Phi¹** offers a nice contrast with the white of nearby **Phi² (φ²)** which lies immediately to the southeast. These two stars are not physically related, Phi¹ lying at a distance of 275 light years, a little over half the distance of Phi² whose light has taken around 520 years to reach us.

Another optical pairing is that formed from **Psi¹ (ψ¹)** and **Psi² (ψ²)**. The yellow, magnitude 4.66 Psi¹ is located at a distance of 219 light years, somewhat

ABOVE: Open star cluster NGC 5822 in Lupus.

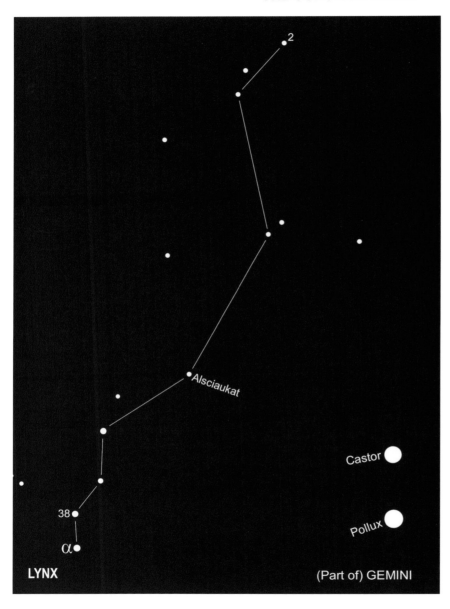

LYNX
The Lynx

The constellation Lynx was introduced in the 17th century by the Polish astronomer Johannes Hevelius in order to fill the gap between Ursa Major to the northeast and Auriga and Gemini to the southwest. The whole of Lynx is visible from latitudes north of 28°S, putting it within the reach of observers in central South America, northern Australia, central

THE STARS OF LYNX

The brightest star in Lynx is **Alpha (α) Lyncis**, a magnitude 3.14 orange giant whose light has taken around 200 years to reach us.

The only named star in this group is **Alsciaukat**, another orange giant star shining at magnitude 4.25 from a distance of nearly 400 light years. The title Alsciaukat is derived from the Arabic for 'Thorn'. An alternative name for this star is Mabsuthat, derived from the Arabic for 'Expanded', referring perhaps to the outstretched paw of the creature depicted by the constellation.

38 Lyncis is a white star shining at magnitude 3.82 from a distance of 125 light years.

Another white star is **2 Lyncis**, the magnitude 4.44 glow of which has reached us from a distance of around 150 light years.

Africa and Indonesia and latitudes to the north of these.

The constellation takes the form of a zigzag line of faint stars which can be picked out by using the two nearby bright stars Castor and Pollux in Gemini as guides. None of the stars in Lynx are brighter than third magnitude although, providing the sky is very dark and clear, you should be able to locate the wandering form of this dim constellation. A pair of binoculars to assist you in your search may be an advantage, testament perhaps to Hevelius not being far off the mark when he explained the constellation name by saying that those who examined the Lynx ought to be lynx-eyed (keen-sighted).

BELOW: Lynx is home to the planetary nebula Jones-Emberson 1. With a diameter of around four light years, Jones-Emberson 1 shines at 14th magnitude from a distance of around 1,600 light years.

LYRA
The Lyre

The small but distinctive Lyra is one of the most prominent constellations in the night sky. Greek legend identifies the group as the lyre given by Apollo to the musician and poet Orpheus to accompany his songs. Its leading star is the brilliant Vega which is visible to observers at mid-northern latitudes at or near the overhead point during evenings in August and September. The whole of Lyra can be seen from latitudes north of 42°S, astronomers located in central South America, South Africa and Australia seeing Vega low down in the northern part of the sky.

GLOBULAR CLUSTER
M56

Lyra plays host to the globular cluster **Messier 56** (M56) or NGC 6779 which shines at magnitude 8.2 from a distance of 32,900 light years. Discovered by Charles Messier in 1779, M56 can be located in binoculars or a small telescope, either of which will show it as a fuzzy patch of light. To find M56, follow the line of faint stars from Sulaphat past 17 and 19 Lyrae and on a little further, as shown on the finder chart. The cluster is seen against the star-filled backdrop of the Milky Way, as alluded to by William Henry Smyth who described it as: *'A globular cluster in a splendid field'*. Take a look at M56 yourself and see if you agree with his observations.

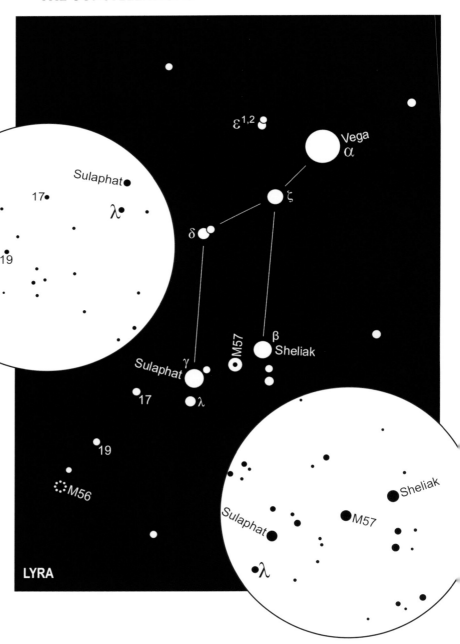

THE STARS OF LYRA

The brilliant blue-white **Vega (α Lyrae)** shines at magnitude 0.03 from a distance of around 25 light years. Vega is the brightest of the three stars that make up the conspicuous asterism known as the Summer Triangle *(see Cygnus)*, the other two being Deneb in Cygnus and Altair in Aquila.

Sulaphat (γ Lyrae) is a magnitude 3.25 white giant star shining from a distance of a little over 600 light years. This star derives its name from the Arabic *'al-sulahfat'* meaning 'the Tortoise'.

Taking its name from the Arabic *'al-salbaq'* meaning 'the Harp', the light from magnitude 3.52 **Sheliak (β Lyrae)** set off towards us around 950 years ago.

Lambda (λ) Lyrae is an orange giant shining at magnitude 4.94 from a distance of around 1,100 light years.

Zeta (ζ) Lyrae is one of several double stars in Lyra, its magnitude 4.34 and 5.73 components easily resolvable in a small telescope or powerful binoculars.

The widely-separated magnitude 4.22 and 5.58 components of the double star **Delta (δ) Lyrae** are clearly resolved in binoculars. The brightest component is a red giant shining from a distance of around 850 light years, the ruddy glow from this star contrasting with the blue-white tint of its companion, the light from which has taken 870 years to reach us.

Immediately to the northwest of Vega is the famous Double-Double star **Epsilon (ε) Lyrae** which, at first glance, appears to be a single star, although observers with exceptionally keen eyesight may be able to identify it as a double. Any form of optical aid will resolve the magnitude 4.59 and 4.62 components, although closer inspection with a large telescope will show that each of these two stars is double again, making Epsilon a quadruple star system.

THE RING NEBULA M57

One of the objects for which Lyra is most famous is the planetary nebula **Messier 57** (M57) or NGC 6720, also known as the Ring Nebula, and discovered in 1779 by the French astronomer Antoine Darquier who described it as: *'A very dull nebula, but perfectly outlined; as large as Jupiter and looks like a fading planet.'* Echoing Darquier's observations of M57 is the fact that many planetary nebulae do indeed have disc-like appearances when viewed through small telescopes.

The description 'planetary nebula' was first coined by William Herschel in 1785 and certainly matches the visual appearance of M57. As is the case with other planetary nebulae, M57 is formed

LEFT: Globular cluster M56 (NGC 6779) in Lyra.

from material thrown off by a star during the latter stages of its evolution, the ejected material having formed a shell around the parent star. As we gaze at M57 we are looking through the nearest wall of the shell of gas, around the edge of which is a more opaque concentration of material which appears as a ring. The well-defined shape of M57, which led to it being christened the Ring Nebula, is not typical of planetary nebulae as a whole, most other known examples having comparatively irregular shapes.

M57 can be seen on a line from Sulaphat to Sheliak, as shown on the finder chart *(see page 140)*, although this object is quite small and faint. Consequently, it may be difficult to see at all unless the sky is really dark, clear and free of moonlight. However, you should take the time to track it down, if only for the fact that you will be gazing at what is perhaps the most famous of the planetary nebulae.

BELOW: The Ring Nebula M57 (NGC 6720) in Lyra.

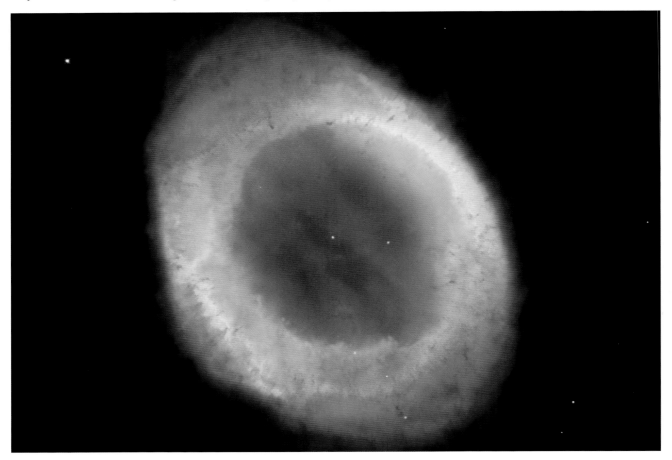

MENSA
The Table Mountain

Visible from regions on or to the south of the equator, this diminutive group is another of the constellations devised by Nicolas Louis de Lacaille and represent the Table Mountain near Cape Town, from where Lacaille charted the southern skies during his stay there in 1751 and 1752. Other than the fact that part of the Large Magellanic Cloud *(see Dorado)* lies within its boundaries, this tiny constellation seems to have little else going for it.

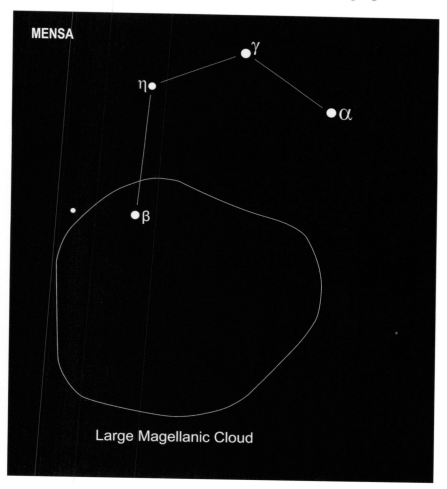

THE STARS OF MENSA

As well as being one of the smallest constellations in the entire sky, Mensa is also the dimmest. Its leading star **Alpha (α) Mensae**, shining at a magnitude of just 5.08 from a distance of 33 light years, is barely visible to the naked eye.

Gamma (γ) Mensae is a magnitude 5.18 orange giant star whose light has taken a little over 100 years to reach us.

Beta (β) Mensae is a magnitude 5.30 yellow giant star located at a distance of around 700 light years.

Completing the main form of Mensa is the magnitude 5.47 orange giant **Eta (η) Mensae**, the light from which set off towards us over 600 years ago.

The tiny circlet of stars comprising Alpha, Gamma, Beta and Eta Mensae which forms the main part of this tiny group can, with a little imagination, be seen to form the shape of a table. Provided the sky is exceptionally dark and clear you should just be able to pick out Mensa with the naked eye by following its curved path away from the southern borders of the Large Magellanic Cloud.

143

MICROSCOPIUM
The Microscope

Taking the form of an elongated rectangle of faint stars, Microscopium lies immediately to the west of Piscis Austrinus and Grus and was one of the constellations representing scientific instruments devised by the French astronomer Nicolas Louis de Lacaille. The whole of Microscopium can be seen from latitudes south of 45°N although none of the stars in the constellation is in any way prominent and none is named.

The chart includes the three nearby stars Alnair and Al Dhanab (both in Grus) and Alpha (α) Indi as guides to locating this faint group.

THE STARS OF MICROSCOPIUM

The brightest star in the constellation is **Gamma (γ) Microscopii**, a yellow giant which shines at a dismal magnitude 4.67 from a distance of around 230 light years.

Epsilon (ε) Microscopii lies just to the east of Gamma and has a magnitude of 4.71, its light having taken around 180 years to reach us.

Alpha (α) Microscopii shines at magnitude 4.89, the light we are seeing from this star arriving from a distance of 375 light years.

Magnitude 4.80 **Theta^1 (θ^1) Microscopii** and magnitude 5.11 **Iota (ι^1) Microscopii** complete the main outline of the group.

MONOCEROS
The Unicorn

Originally devised by the Dutch celestial cartographer Petrus Plancius and depicted on a globe produced by him in 1613, the constellation Monoceros depicts the legendary animal with the body of a horse and a single horn in the middle of its forehead. Monoceros lies across the celestial equator, occupying the area of sky between Orion, Canis Major, Canis Minor and Hydra, and is visible in its entirety from virtually anywhere on Earth.

None of the stars that make up this constellation are particularly bright, although locating Monoceros is fairly straightforward as the bulk of it lies within the so-called Winter Triangle, this being an asterism formed by the three bright stars Betelgeuse (in Orion), Procyon (in Canis Minor) and Sirius (in Canis Major).

OPEN STAR CLUSTER M50
The open star cluster **Messier 50** (M50) or NGC 2323 is located near the southern border of the constellation, roughly half way between Alpha (α) and Beta (β). M50 shines with an overall magnitude of 5.9 and may just be visible to the naked eye under an exceptionally dark, clear sky. The English astronomer John Herschel described M50 as: *'A remarkable cluster, very large and rich . . .'* and, as his

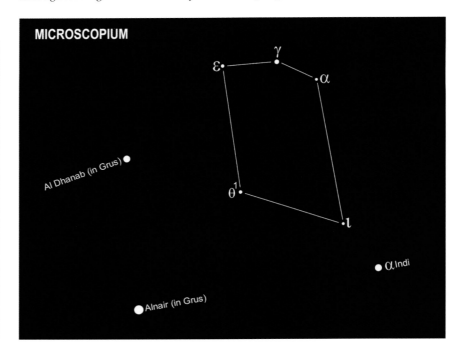

MICROSCOPIUM

THE STARS OF MONOCEROS

Although not the most conspicuous star in the constellation, **Epsilon (ε) Monocerotis** is a good place to start, being located at the northwestern end of Monoceros, not far from Betelgeuse in the neighboring constellation Orion. Shining at magnitude 4.39 Epsilon can be found by following a line slightly south of eastwards from Betelgeuse. Once you locate Epsilon you can trace out the rest of Monoceros as it trails off to eastwards by using the chart as a guide and a pair of binoculars to make the search easier. Because the stars in this constellation are all relatively faint, you'll need a clear, dark and moonless sky to see them with the naked eye.

The brightest star in Monoceros is **Beta (β) Monocerotis** which shines at magnitude 3.76 from a distance of around 700 light years. Beta is actually a triple star system, discovered by William Herschel in 1781 and described by him as: *'One of the most beautiful sights in the heavens'*. Beta comprises a dazzling gravitationally-linked trio of components designated β Monocerotis A, β Monocerotis B and β Monocerotis C. When viewed through a small telescope, magnitude 5.40 A forms a pretty double with B and C which, because they are so close together, appear as a single star. However, a slightly larger telescope will split the magnitude 5.40 and 5.60 B and C pair, the three components then forming a lovely curved row of stars.

Alpha (α) Monocerotis is a magnitude 3.94 orange giant, the light from which has taken almost 150 years to reach us.

Another orange giant star is magnitude 3.99 **Gamma (γ) Monocerotis** which shines from a distance of around 500 light years.

Zeta (ζ) Monocerotis can be found at the eastern end of the constellation glowing at magnitude 4.36 from a distance of around 1,000 light years.

The light from magnitude 4.15 **Delta (δ) Monocerotis** set off towards us around 380 years ago.

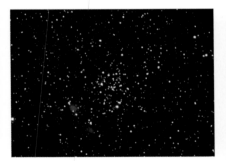

description suggests, the cluster is fairly easy to find. To track it down, use binoculars to follow a line from Alpha working westwards towards Beta and you will eventually reach M50. The cluster will be visible as a misty patch of light against

ABOVE: Open star cluster M50 (NGC 2323) in Monoceros.

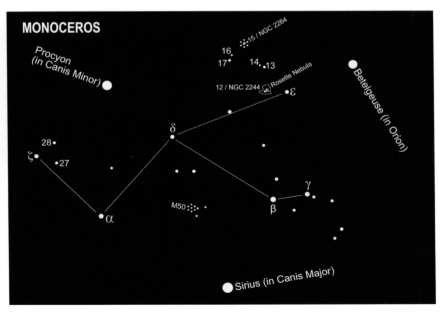

the background stars, located immediately to the east of the small triangle of stars shown here. M50 lies at a distance of around 3,200 light years and contains anything up to 200 stars, a number of which should be revealed in large binoculars or a small telescope.

OPEN STAR CLUSTER NGC 2244

Located at a distance of over 5,000 light years, and discovered by the English astronomer John Flamsteed in around 1690, is the open star cluster **NGC 2244**. Situated just to the east of Epsilon, this cluster lies within the same binocular field of view and forms a small triangle with Epsilon and the nearby star 13 Monocerotis. NGC 2244 surrounds the star 12 Monocerotis and can be tracked down fairly easily in binoculars.

THE ROSETTE NEBULA

The Rosette Nebula is a large ring of nebulosity surrounding the NGC 2244 cluster. The Rosette is an example of an emission nebula *(see Glossary)* in that it shines through the effects of young, hot stars embedded within it. When photographed with telescopes, this object is one of the most attractive of the nebulae, although it is extremely difficult to observe directly. However, providing the sky is very dark, clear and free of moonlight, good binoculars or a small

OPPOSITE: The open star cluster NGC 2244 with the Rosette Nebula.

RIGHT: The Christmas Tree Cluster (NGC 2264) in Monoceros.

telescope may reveal the Rosette Nebula as a patch of very soft light surrounding the NGC 2244 cluster. If you have problems seeing it, averted vision *(see Glossary)* may be the key to catching a glimpse of this visually-elusive object.

THE CHRISTMAS TREE CLUSTER

The open star cluster NGC 2264 lies at a distance of around 2,400 light years and, shining with an overall magnitude of 3.9, is an easy target for binoculars. Also known as the Christmas Tree Cluster, due

BELOW: The appropriately-named Cone Nebula is seen here to the right of the image. Discovered by William Herschel in 1785, the Cone Nebula appears as a notch in the southern end of the nebulosity which surrounds the Christmas Tree Cluster, which is also evident in this picture.

to its visual appearance when viewed through telescopes, NGC 2264 may be glimpsed with the naked eye under dark, clear skies. Containing around 30 member stars, this is another of the large number of open clusters found in Monoceros. Star 15, located a little way to the northeast of 13, lies on the edge of the cluster.

BELOW: Seen against a backdrop of stars, NGC 2170 is a wonderful mixture of blue reflection, red emission and dark nebulosity in Monoceros.

MUSCA
The Fly

One of the groups introduced by the Dutch navigators and explorers Pieter Dirkszoon Keyser and Frederick de Houtman, following their expedition to the East Indies in the 1590s, the tiny constellation Musca can be found immediately to the south of Crux. The diminutive but conspicuous form of Musca can only be readily seen from locations on or south of the equator and is completely hidden to observers north of latitude 25°N. The chart shows Musca together with the two nearby prominent stars Acrux in Crux and Beta (β) Centauri in Centaurus which will help you to track this constellation down.

Originally known as Apis (the Bee), this constellation was included on the celestial globe produced by the Dutch celestial cartographer Petrus Plancius in 1598, as well as in the *Uranometria*, a star atlas produced by the German astronomer and celestial cartographer Johann Bayer in 1603. The French astronomer Nicolas Louis de Lacaille renamed the constellation Musca Australis (the Southern Fly), avoiding any confusion with Musca Borealis (the Northern Fly), a constellation which existed at the time immediately to the northeast of the constellation Aries (Aries). When Musca Borealis lost its place on star charts, the only fly that astronomers were left with was Musca Australis, the name of which was subsequently reduced to Musca.

OPPOSITE: The spiral planetary nebula NGC 5189 in Musca.

THE STARS OF MUSCA

The two brightest stars in Musca are **Alpha (α) Muscae** and **Beta (β) Muscae**. Alpha shines at magnitude 2.69 from a distance of 315 light years, whilst Beta is a little further away, the light from this magnitude 3.04 star having taken 340 years to reach us.

Shining at magnitude 3.84, **Gamma (γ) Muscae** lies at a distance of around 325 light years

Delta (δ) Muscae completes the quadrilateral forming the main section of Musca, this magnitude 3.61 orange giant shining from a distance of 91 light years.

The light from magnitude 3.63 **Lambda (λ) Muscae**, located near the northwestern corner of Musca, set off towards us around 125 years ago.

Backyard astronomers equipped with a small telescope might like to check out the yellowish and white magnitude 5.7 and 7.3 components of the double star **Theta (θ) Muscae**, which lies almost on a line from Alpha through Beta Muscae. Musca lies on the edge of the Milky Way and Theta forms a pretty sight when viewed against the rich field of stars that act as a backdrop to this pair.

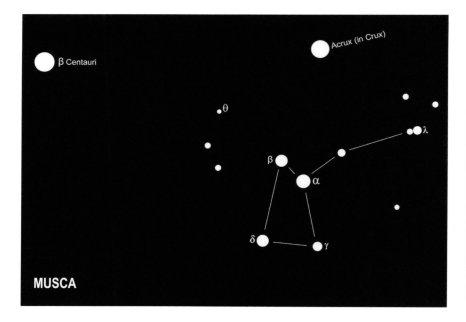

MUSCA

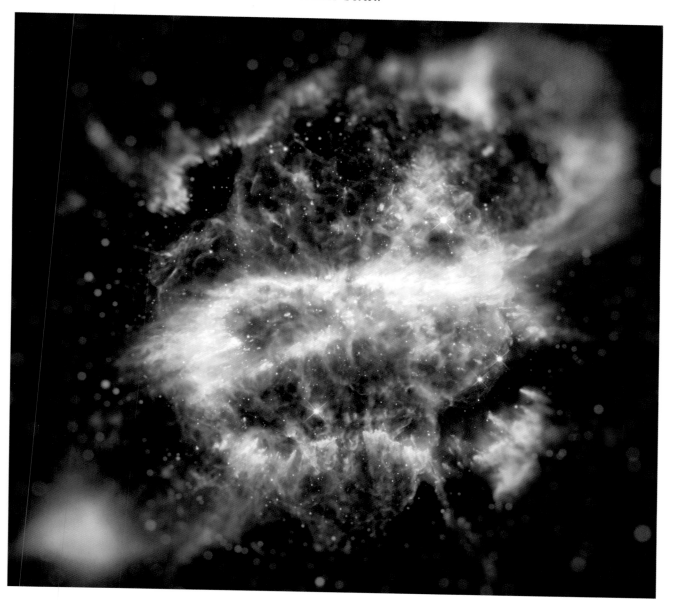

NORMA
The Level

Devised by the French astronomer Nicolas Louis de Lacaille to represent the set square used by draftsmen, the constellation Norma takes the form of a tiny quadrilateral of faint stars located a little way to the northeast of the prominent pair Alpha (α) and Beta (β) Centauri, which are included on the chart as guides to locating this group. Norma can be viewed in its entirety from anywhere south of latitude 30°N although, as with many of Lacaille's

THE STARS OF NORMA

The brightest star in Norma is **Gamma^2 (γ^2) Normae**, a magnitude 4.01 orange giant shining from a distance of around 130 light years. The magnitude 4.97 white supergiant **Gamma^1 (γ^1) Normae** lies close to Gamma^2 although the two stars are not related. The pair form an optical double, measurements revealing that Gamma^1 shines from a distance of well over 1,000 light years.

Delta (δ) Normae shines at magnitude 4.73 from a distance of around 120 light years.

The light from magnitude 4.65 yellow giant **Eta (η) Normae** set off on its journey towards us a little over 200 years ago.

Epsilon (ε) Normae completes the main quadrilateral of Norma and is a double star resolvable in small telescopes. Located at a distance of around 530 light years, the main component of Epsilon is blue and shines at magnitude 4.8, although a small telescope or powerful binoculars will reveal a magnitude 7.5 companion star.

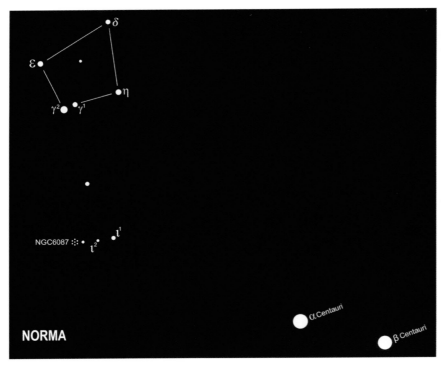

creations, none of the stars in the constellation is named and none is particularly prominent.

OPEN STAR CLUSTER NGC 6087

Located a little way to the east of the stars Iota[1] (ι^1) Normae and Iota[2] (ι^2) Normae is the magnitude 5.4 open star cluster **NGC 6087** which contains around 40 member stars and shines from a distance of around 3,500 light years. This is a fairly easy binocular object although a telescope will be needed to resolve any individual stars in the cluster.

OPPOSITE: Open star cluster NGC 6087 in Norma.

BELOW: A section of the Norma Cluster, a rich cluster of galaxies in Norma which is located at a distance of over 200 million light years.

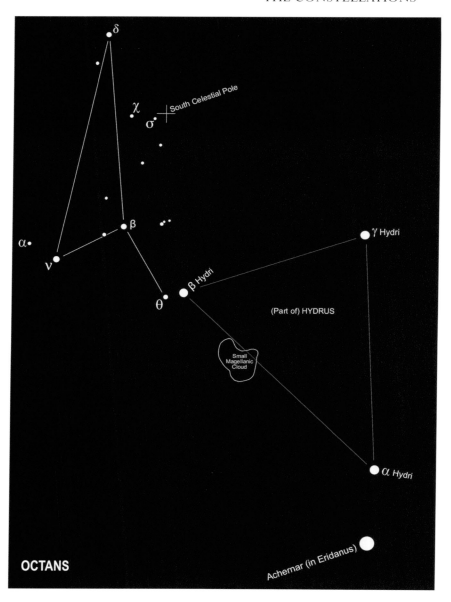

South Celestial Pole

(Part of) HYDRUS

Small Magellanic Cloud

γ Hydri

β Hydri

α Hydri

Achernar (in Eridanus)

OCTANS

OCTANS
The Octant

Another of the groups introduced by the French astronomer Nicolas Louis de Lacaille in the 1750s, Octans represents the navigational instrument invented by the English mathematician John Hadley to measure the altitude of stars and other celestial objects above the horizon. With this in mind, it is appropriate that the south celestial pole *(see left)* lies within its boundaries. As is the case with many of the constellations devised by Lacaille, Octans is somewhat obscure and contains no particularly bright stars.

Because Octans is situated close to the south celestial pole, it can only be seen in its entirety by stargazers located on or south of the equator. To find Octans, first of all locate the bright star Achernar in Eridanus, just to the south of which you will see Alpha (α) Hydri, one of the brighter members of the neighboring constellation Hydrus. Alpha, Gamma (γ) and Beta (β) Hydri form the main outline of Hydrus, the star Theta (θ) Octantis lying close to Beta Hydri. Once you have identified Theta, the rest of Octans can be picked out, Achernar and the main stars forming Hydrus being depicted on the chart to help you track down and identify this faint group.

OPPOSITE: Located immediately to the south of the star Rho (ρ) Ophiuchi is the Rho Ophiuchi cloud complex, the prominent region of gas and dust seen here at right of picture. The Milky Way crosses the left section of this wide-field image.

THE STARS OF OCTANS

Alpha (α) Octantis is a white giant star shining at magnitude 5.13 from a distance of around 140 light years.

Situated just a short distance from Alpha is **Nu (ν) Octantis**, the brightest star in Octans. The light from this magnitude 3.73 orange giant set off on its journey towards us 69 years ago.

Theta (θ) Octantis is another orange giant, its magnitude 4.78 glow reaching us from a distance of around 220 light years

Yet another orange giant is **Delta (δ) Octantis**, the light from this magnitude 4.31 star having taken almost 300 years to reach us.

Completing the main outline of Octans is **Beta (β) Octantis**, a magnitude 4.13 star located at a distance of around 150 light years.

The north (and south) celestial poles are the points on the celestial sphere through which extensions of the Earth's axis of rotation would pass *(see Glossary)*. The north celestial pole lies within the constellation Ursa Minor and is marked by the star Polaris *(see Ursa Minor)*, the southern equivalent of Polaris being **Sigma (σ) Octantis** which lies only a short distance from the south celestial pole. While Polaris is comparatively bright and can be picked out fairly easily, Sigma Octantis shines at a feeble magnitude 5.46 and is barely visible to the naked eye.

To locate Sigma, first of all pick out the main shape of Octans, then follow the line from Beta towards Delta. The two faint stars Chi (χ) and Sigma Octantis will be seen near this line. Unless the sky is really dark and clear you may need some form of optical aid to see Sigma, and once you do manage to locate this faint star, you'll realize how fortunate those in the northern hemisphere are to have such a relatively bright pole star as Polaris!

OPHIUCHUS
The Serpent Bearer

The constellation Ophiuchus depicts a man grasping a huge snake (represented by the adjoining constellation Serpens) with the head of the snake in his left hand and the tail in his right. Both Ophiuchus and Serpens were among the 48 constellations listed by the Greek astronomer Ptolemy during the second century and together they straddle the celestial equator, the result of which is that the group can be seen in its entirety from all inhabited regions south of latitude 60°N, with portions being visible worldwide. According to legend, Ophiuchus represents Asclepius, the god of medicine. Because snakes were considered to be a symbol of healing, it is perhaps fitting that Asclepius is depicted as holding a snake.

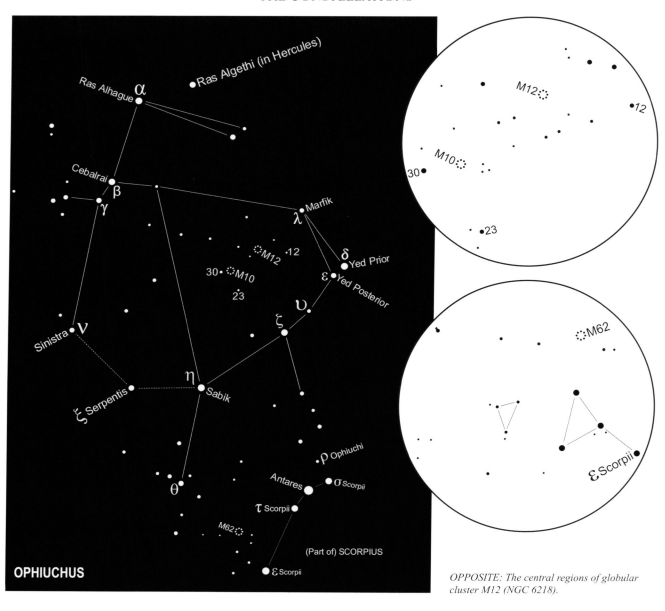

Ras Algethi (in Hercules)

Ras Alhague α

Cebalrai β
γ

Marfik λ

δ Yed Prior

M12
·12

30· M10

23

ε Yed Posterior

υ

ζ

Sinistra ν

η Sabik

ξ Serpentis

θ

ρ Ophiuchi

Antares
σ Scorpii

τ Scorpii

M62

(Part of) SCORPIUS

ε Scorpii

OPHIUCHUS

M12
·12

M10
30

23

M62

ε Scorpii

OPPOSITE: The central regions of globular cluster M12 (NGC 6218).

GLOBULAR CLUSTERS M10 AND M12

Amongst the large number of globular clusters in Ophiuchus are **Messier 10 (M10)** or NGC 6254 and **Messier 12 (M12)** or NGC 6218, both of which are located to the east of Yed Prior and Yed Posterior. Discovered by Charles Messier in 1764, M10 glows at magnitude 6.4 from a distance of around 15,000 light years and lies in the same binocular field of view as nearby M12. Messier described M10 as a *'Nebula, without star in the belt of Ophiuchus'*, the cluster having first been resolved into stars by William Herschel, who saw it as *'A beautiful cluster of extremely compressed stars . . .'*.

Immediately to the northwest of M10 we find M12. This globular cluster was also discovered by Charles Messier in 1764, his description of it being a *' . . . nebula (which) contains no star, round, faint . . .'*, echoing his visual impression of nearby M10. As was the case with M10, William Herschel's superior telescope allowed him to resolve M12 into

THE STARS OF OPHIUCHUS

Ophiuchus is shown here along with a number of guide stars to help you pick out the main outline of the constellation. These include Xi (ξ) Serpentis in Serpens to the southeast and, to the southwest, a section of the bordering constellation Scorpius, together with the brilliant Antares. The prominent star Ras Algethi in Hercules lies immediately to the north of Ophiuchus, and a little to the east of Ras Algethi we find **Ras Alhague (α Ophiuchi)**, the brightest star in the constellation. Ras Alhague takes its name from the Arabic for 'the Head of the Serpent Collector' and shines at magnitude 2.08 from a distance of 47 light years.

Sabik (η Ophiuchi) is the second brightest star in Ophiuchus, its magnitude 2.43 glow reaching us from a distance of 88 light years. Slightly fainter is magnitude 2.54 **Zeta (ζ) Ophiuchi**, the light from which set off towards us over 350 years ago.

Upsilon (υ) Ophiuchi shines at magnitude 4.62 from a distance of 122 light years.

Deriving their unusual names from the Arabic *'al-yad'* meaning 'the Hand', the pair of stars **Yed Prior (δ Ophiuchi)** and **Yed Posterior (ε Ophiuchi)** are located near the western border of Ophiuchus, Yed Prior representing the foremost hand and Yed Posterior the hindmost. The red giant Yed

Prior shines at magnitude 2.73 from a distance of 170 light years, nearly twice the distance of the magnitude 3.23 yellow giant Yed Posterior, the light from which reaches us from a little over 100 light years away.

Located to the northeast of Yed Prior and Yed Posterior, and shining at magnitude 3.82 from a distance of around 170 light years, the blue giant **Marfik (λ Ophiuchi)** derives its name from the Arabic *'al-marfiq'* meaning 'the Elbow'.

The magnitude 2.76 orange giant star **Cebalrai (β Ophiuchi)** lies a little to the south of Ras Alhague. Shining from a distance of 82 light years, Cebalrai takes its name from the Arabic *'kalb al-ra'i'* meaning 'the Shepherd's Dog'.

Slightly more remote than Cebalrai is magnitude 3.75 **Gamma (γ) Ophiuchi**, the light from which has taken 103 years to reach us.

Located near the southern border of Ophiuchus is **Theta (θ) Ophiuchi**, shining at magnitude 3.27 from a distance of around 435 light years.

The light from magnitude 3.32 orange giant **Sinistra (ν Ophiuchi)** set off on its journey towards us 151 years ago.

individual stars, Herschel describing it as *'A brilliant cluster . . .'*.

Located at a distance of over 15,000 light years, M12 shines at magnitude 6.7 and is slightly fainter than M10. Both objects can be tracked down in binoculars, providing the sky is dark and clear. Start your search by checking out the area to

the east of Yed Prior and Yed Posterior and locating the triangle of faint stars formed from 30, 12 and 23 Ophiuchi. You can then use the detailed finder chart *(page 156)* to star-hop your way to M10 through the field of faint stars in which these two clusters lie. As is the case with most objects of this type, remember to look for a patch of light rather than a star-like point. Neither M10 nor M12 is particularly prominent so patience may be needed in your efforts to track these two objects down.

GLOBULAR CLUSTER M62

Of the other globular clusters in Ophiuchus, perhaps the best example is **Messier 62** (M62) or NGC 6266 which lies on the southern border of Ophiuchus to the east of Antares in the neighboring constellation Scorpius. Observers at mid-northern latitudes will not find this object easy to track down as it lies fairly close to the horizon, although it will be more accessible to those further south.

Discovered by Messier in 1771 and first resolved into stars by William Herschel, M62 shines at around magnitude 6.5 from a distance of 22,500 light years. Herschel was inspired by his view of M62, describing it as '. . . a *miniature of M3'*, (M3 is a particularly fine globular cluster in the constellation Canes Venatici).

Although M62 is well within the light grasp of binoculars you will need at least a

OPPOSITE: A Hubble Space Telescope image capturing the central region of globular cluster M10 (NGC 6254).

RIGHT: Globular cluster M62 (NGC 6266).

small telescope in order to resolve individual stars within the cluster, although binoculars will reveal M62 as a small nebulous object which will be decidedly non-stellar in appearance. Start your search for M62 by locating the guide star Epsilon (ε) Scorpii to the southeast of Antares in neighboring Scorpius. This star is also shown on the finder chart *(page 156)* and, once you have identified it, you can use binoculars to work your way towards M62 through the pattern of faint stars shown. The two tiny triangles situated immediately to the east of Epsilon are easily picked up and these can be used as a guide to take you to M62.

ORION
Orion

The conspicuous pattern of bright stars forming the constellation Orion is unmistakable. Visible in its entirety from almost every inhabited part of the world, it straddles the celestial equator and is visible high in the southern winter sky (for northern hemisphere observers) and high in the northern summer sky (for observers in the southern hemisphere). This magnificent collection of stars is arguably the most beautiful constellation in the heavens.

According to legend, Orion was one of the sons of Neptune, which in itself would be something to be proud of. However, Orion's vanity went a stage further and he boasted far and wide of his prowess in the field, so much so that he declared himself to be a match even for the gods. Diana, the Goddess of the Hunt, was somewhat put out by this and, in an effort to bring Orion down a peg or two, challenged Orion to a hunting match. Orion was delighted for an opportunity to boost his self esteem even further and gladly accepted the challenge. Many days and nights were spent in the chase, but in the end there was nothing to choose between them, and the contest was declared a draw. Naturally, this did little to curb Orion's vanity and, in a fit of anger, Diana commanded a scorpion to crawl out of the ground and kill Orion, following which she placed the mighty hunter among the stars where he can be seen to this day.

The main shape of the constellation is outlined by Betelgeuse, Bellatrix, Rigel

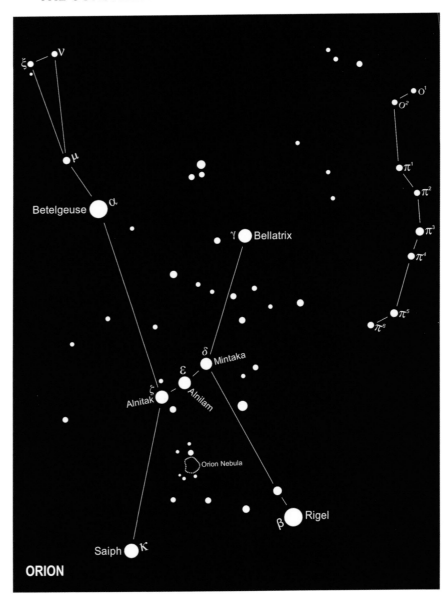

THE STARS OF ORION

Betelgeuse (α Orionis) marks the northeastern corner of Orion. The name of this brilliant red supergiant star is derived from the Arabic **'Yad al-jauza'** meaning 'the Hand of al-jauza'. Medieval translators mistakenly took the first letter as being a 'B' rather than a 'Y', from which the currently accepted name of this star eventually emerged. The exact identity of 'al-jauza' is not clear, although it appears to be a reference to a female figure that Arabic astronomers saw as being depicted by the stars that represent Orion.

Shining from a distance in excess of 500 light years, Betelgeuse is one of the largest stars known to astronomers. Estimates of its diameter are such that, if Betelgeuse occupied the position of our Sun, its outer surface would extend beyond the orbit of Mars. Betelgeuse pulsates slightly, resulting in slight changes to its diameter and consequent variations in its brightness. However, taking place over periods of several years, these changes are only very slight and not particularly noticeable.

Located at the southwestern corner of Orion we find the brilliant star **Rigel (β Orionis)**. Its name derived from the Arabic **'rijl'** meaning 'Foot', Rigel is another supergiant star with a diameter of around 40 million miles and a luminosity of over 50,000 times that of our Sun. The difference between the ruddy glow of Betelgeuse and the brilliant white of Rigel is evident even when viewed with the naked eye and checking them out with binoculars brings out the color difference between them very well. Rigel lies at a distance of around 850 light years, which means the light we are seeing now set off on its immense journey towards us not long after the Domesday Book was completed in 1086.

The third brightest star in Orion is **Bellatrix (γ Orionis)** which shines at magnitude 1.64 from a distance of around 250 light years from where it depicts the left arm of Orion. Otherwise known as the Amazon Star, Bellatrix takes its name from the Latin for 'the Female Warrior'.

Saiph (κ Orionis) is the sixth brightest star in the constellation with a magnitude of 2.07, its light having set off on its journey towards us around 650 years ago. Marking Orion's right knee, its name is derived from the Arabic '**saif al-jabbar'** meaning 'the Sword of the Giant.'

The easternmost of the three stars that form the Belt of Orion is **Alnitak (ζ Orionis)**. Shining at magnitude 1.75 from a distance of around 1,000 light years, its name is derived from the Arabic for 'the Girdle'. The area immediately surrounding Alnitak is rich in stars and well worth a look through binoculars.

Alnilam (ε Orionis) shines at magnitude 1.70 from a distance of over 1,300 light years. Its name is derived from the Arabic **'al-nizam'** meaning 'the String of Pearls'.

Shining at magnitude 2.25, its light having taken around 1,000 years to reach us, **Mintaka (δ Orionis)** is located immediately to the west of Alnilam, completing the Belt of Orion.

Although we identify the trio of stars Alnitak, Alnilam and Mintaka as depicting the Belt of Orion, they have collectively been known by other names in the past including 'Jacob's Rod' and 'The Three Kings'. Greenlanders referred to them as 'The Seal Hunters' and mariners often identified them as 'The Golden Yardarm'. However, the most ridiculous episode relating to the naming of the three stars has to be that which occurred in 1807, when an irate Englishman gave them the title 'Nelson'. This was in response to the University of Leipzig which had christened them 'Napoleon'. Happily, neither name actually found its way onto star maps!

and Saiph which together form the conspicuous Rectangle of Orion. If the night is really dark and clear you might be able to make out the line of faint stars extending from **Omicron[1] (o[1])** and **Omicron[2] (o[2])** southwards to **Pi[6] (π[6])**. This represents the lion's skin which Orion uses as a shield. If you have problems picking these stars out with the naked eye, a pair of binoculars will help. Dark skies should also reveal Orion's club, which is depicted by the triangle of faint stars **Mu (μ)**, **Nu (μ)** and **Xi (ξ)** and located immediately to the north of Betelgeuse.

THE GREAT ORION NEBULA M42

A little way to the south of the Belt of Orion we see a line of faint stars which are commonly referred to as the Sword of Orion and which, if the night is dark and clear, you should be able to make out with the unaided eye. Along with these you might glimpse a hazy and diffuse patch of light surrounding the group of stars at the southern end of the Sword. Known as the Great Orion Nebula, this is the most famous object in the constellation and is by far Orion's main claim to fame.

Also known as **Messier 42** (M42) or NGC 1976, the Orion Nebula takes the form of a giant, irregular glowing cloud. Shining because of the stars embedded within it, this magnificent object lies at a

BELOW: The Great Orion Nebula M42 (NGC 1976).

OPPOSITE: Located immediately to the northeast of M42 is the mixture of emission and reflection nebulae NGC 1977, also known as the Running Man Nebula.

distance of around 1,400 light years and has a diameter of at least 25 light years. On really dark, clear and moonless nights it can be seen as a faint glowing patch of light and it is remarkable that its existence appears not to have been noted until 1611. Since then, it has never failed to impress, and it is safe to say that there is little in the heavens to rival it. The wide field of view of a pair of binoculars brings out the nebula very well, and the sight leaves the observer with a sense of wonder when it is realized that stars are actually being formed inside this glowing cloud.

The beautiful constellation of Orion seldom fails to impress, a sentiment echoed in the words of the English astronomer Joseph Henry Elgie who said of it: *'Orion at midnight strode the southern sky like a Colossus. What an attention-compelling constellation Orion is! With its heroic proportions, its conspicuous stars, Betelgeuse, Rigel and Bellatrix; its jewel-sparkling Belt; its pendant Sword; and its stupendous nebula, it excites the highest admiration of astronomer and casual observer alike.'* The phrase 'attention-compelling' and other flattery would certainly suit the vanity of the mighty Orion himself!

LEFT: The emission nebula known as Barnard's Loop is seen here alongside the major stars of Orion.

OPPOSITE: A region of star formation in the constellation Orion, photographed in infrared by the Spitzer Space Telescope.

PAVO

The Peacock

Pavo takes the form of an extended oval of stars located to the east of the constellation Ara, its westernmost star Eta (η) Pavonis lying a little to the southeast of the three stars Beta (β), Gamma (γ) and Delta (δ) Arae, all three of which are depicted on the chart as a guide.

This is another of the groups introduced following observations made by Pieter Dirkszoon Keyser and Frederick de Houtman during the 1590s and depicts a peacock, a bird which was somewhat limited in distribution at the time the constellation was devised but which is now found in parks and gardens throughout the world. Pavo can be seen in its entirety from latitudes south of 15°N.

THE GREAT PEACOCK CLUSTER

The Great Peacock Cluster, or NGC 6752, is considered to be one of the finest globular clusters in the sky. Discovered by the Scottish astronomer James Dunlop in 1826 this fine object lies at a distance of around 13,000 light years. Shining at

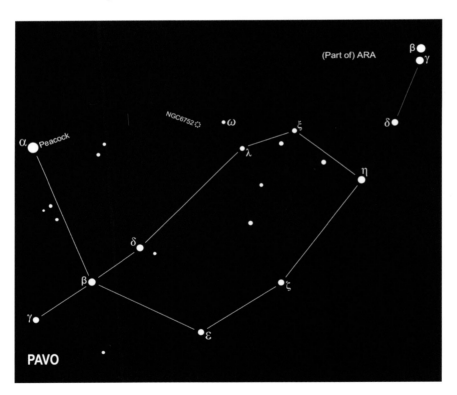

(Part of) ARA

β
γ

NGC6752

ω

ξ

δ

α Peacock

λ

η

δ

β

ζ

γ

ε

PAVO

around magnitude 5.4, NGC 6752 can be glimpsed with the naked eye a little way to the east of the star Omega (ω) Pavonis. Binoculars bring it out well and even a small telescope will resolve individual stars in its outer regions. The Great Peacock Cluster is a must-see for the backyard astronomer!

BELOW: The Great Peacock Cluster (NGC 6752) in Pavo.

PEGASUS
The Winged Horse

For observers in the northern hemisphere, Pegasus is the main autumnal constellation although, being located a little to the north of the celestial equator, it is fairly easy to locate from almost anywhere. Pegasus is the 7th largest constellation although, in spite of covering an extensive area of sky, it contains no really bright stars. The Square of Pegasus is quite distinctive, however, and it is an interesting exercise to try to count the number of naked-eye stars you can detect within its boundaries. Observers with really keen eyesight may be able to spot around two dozen, although you will need a really clear, dark sky for your search.

The constellation is best placed for observation around October, at which time backyard astronomers at mid-northern latitudes will see Pegasus riding high above the southern horizon, with observers in the southern hemisphere seeing the Winged Horse gracing their northern skies. The entire constellation can be seen from latitudes north of 54°S and, once the Square of Pegasus has been identified, the rest of the group can be picked out trailing away towards the west. It should be pointed out that Sirrah, the star at the northeastern corner of the Square, is actually a member of the neighboring constellation Andromeda, from where it is 'borrowed' to complete the Square of Pegasus on star charts.

The constellation represents Pegasus, the son of Neptune and Medusa. When Perseus slew Medusa, Pegasus sprang from her decapitated body and flew away, eventually reaching Corinth. Here he was found by Bellerophon, who tamed Pegasus and used him in his fight against the fire-breathing monster Chimaera. After many adventures together, Bellerophon attempted to fly to Olympus, the home of the gods, although Pegasus threw Bellerophon on the way and completed the journey alone. Zeus, the ruler of Olympus, placed the winged horse in the sky where we see him today.

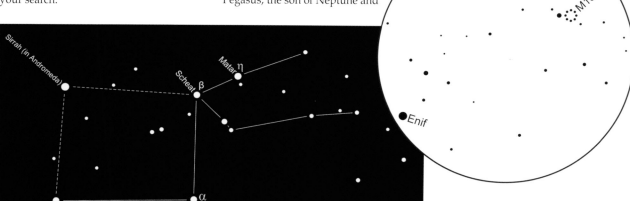

GLOBULAR CLUSTER M15

Discovered by the Italian astronomer Jean Dominique Maraldi in 1746, and situated just to the northwest of Enif, is the beautiful globular star cluster Messier 15 (M15) or NGC 7078. Located at a distance of over 33,000 light years, this object was

THE STARS OF PEGASUS

Algenib (γ Pegasi) marks the southeastern corner of the Square. Deriving its name from the Arabic *'al-janb'* meaning 'wing' or 'side' Algenib shines at magnitude 2.83 from a distance of around 390 light years.

The southwestern corner of the Square is depicted by magnitude 2.49 **Markab (α Pegasi)**, the light from which has taken around 130 years to reach us. The name of this star is derived from the Arabic *'mankib al-faras'* meaning 'the Horse's Shoulder'.

The light from the magnitude 2.44 red giant **Scheat (β Pegasi)**, located at the northwestern corner of the Square, set off on its journey towards us nearly 200 years ago. Scheat takes its name from the '*Arabic 'al-saq'* meaning 'the Shin'.

Shining with a magnitude of 2.38 from a distance of just under 700 light years, the orange supergiant **Enif (ε Pegasi)** is the brightest star in Pegasus and denotes the horse's head. This star may derive its name from the Arabic *'anf'* meaning 'nose', although Arabic astronomers themselves occasionally referred to Enif as the horse's mouth.

Homam (ζ Pegasi) shines at magnitude 3.41 from a distance of a little over 200 light years.

Matar (η Pegasi) is a yellow giant star, the magnitude 2.93 glow of which reaches us from a distance of around 215 light years.

The magnitude 3.52 glow of **Biham (θ Pegasi)**, located close to the southern border of Pegasus, reaches us from a distance of 92 light years.

described by Maraldi as: *'A nebulous star, fairly bright and composed of many stars'.*

M15 lies just below the threshold of naked-eye visibility, although its overall magnitude of 6.2 makes it an easy target for binoculars or a small telescope. To locate the cluster, first of all find the bright star Enif, from where you can star hop your way northwestwards to M15 through the field of fainter stars shown on the finder chart. Providing the sky is dark and clear, M15 is bright enough to be detected by carefully sweeping the area with a pair of binoculars. Once found, the cluster should appear as a small diffuse patch of light, telescopes being required to resolve any individual stars.

RIGHT: Globular cluster M15 (NGC 7078) in Pegasus.

PERSEUS
Perseus

Depicting the legendary Greek hero of the same name, the constellation Perseus is located to the east of Andromeda, the star Almach in Andromeda included on the chart as a guide to locating this famous group. The constellation is visible in its entirety from locations to the north of 31°S, and backyard astronomers in northern Australia, South Africa and central South America will see most of the constellation suspended low over the northern horizon. Portions of the constellation will be permanently hidden from the view of observers at latitudes south of these.

Perseus was the son of Zeus and Danaë, his main claims to fame being the slaying of Medusa, one of the three Gorgons, and the rescue of (and subsequent marriage to) Andromeda, the daughter of King Cepheus and Queen

Cassiopeia of Ethiopia. Following a long, prosperous and adventure-filled life Perseus died and was placed among the stars alongside his wife Andromeda and his in-laws Cepheus and Cassiopeia. The sea monster sent to kill Andromeda is also there, albeit somewhat further to the south, represented by the constellation Cetus. To complete the scenario, the head of Medusa is depicted by the star Algol.

ALGOL

Deriving its name from the Arabic *'ra's al-ghul'* meaning 'the Demon's Head' and, according to legend, depicting the severed head of Medusa the Gorgon, held by Perseus, **Algol (β Persei)** is one of the most famous variable stars in the entire sky and, located at a distance of just 93 light years, the closest object of its kind.

Algol is what astronomers refer to as an eclipsing binary, its changes in brightness being due to the intervention of another star rather than through changes taking place within the star itself. There are many binary star systems in the sky, each with components that are in orbit around each other, and it is reasonable to assume that the orbital planes of many of these will be lined up with our position in space. In the systems where this situation occurs, such as Algol, the overall light output of the binary decreases as the fainter of the two stars crosses our line of sight, passing in front of the brighter component and temporarily obscuring some of the light from the brighter star. There is, of course, a second dip in the light output of eclipsing binaries, this taking place when the fainter star passes behind the brighter component, although

the resulting reductions in brightness are generally nowhere near as great.

Algol is just one example of several thousand eclipsing binary stars known to astronomers. In the case of Algol, one star is considerably fainter than its companion and the plane of their orbit is almost exactly lined up with our position in space. Consequently, when the main eclipses occur, the overall light output from Algol decreases markedly from magnitude 2.1 to 3.4 before climbing again. The entire sequence takes around 10 hours with a well-determined period between successive times of minimum brightness of 2.867 days (2 days, 20 hours, 48 minutes, 56 seconds).

Algol's changes in brightness may have been known as long ago as the tenth century, although the earliest recorded observations of its variability were made in 1667 by the Italian astronomer Geminiano Montanari, who happened to notice that Algol was shining at less than its usual brightness. He was inspired enough to make a series of observations which led him to the conclusion that Algol was indeed variable. However, it was the astronomer John Goodricke who made the first accurate measurements of its period of variability. It was also Goodricke who first suggested that Algol was a binary and that the variations in brightness were due to the occasional eclipse of a brighter star by a fainter companion.

By making a series of random checks on Algol over several nights, you should eventually be able to detect its fluctuations, from which point you can follow its changes in magnitude by comparing it to nearby stars. However,

THE STARS OF PERSEUS

The brightest star in Perseus is the white supergiant **Algenib (α Persei)**. Shining at magnitude 1.79 from a distance of around 580 light years, Algenib derives its name from the *Arabic 'al-janb'* meaning 'the Side or Flank'.

Atik (o Persei) is a magnitude 3.84 blue supergiant whose light has taken over 850 years to reach our planet.

Menkib (ξ Persei) lies a little to the north of Atik, the light from this magnitude 3.98 star having set off towards us around 1,200 years ago.

The magnitude 2.91 yellow giant star **Gamma (γ) Persei** shines from a distance of around 250 light years.

The light from magnitude 3.77 orange giant star **Miram (η Persei)** reaches us from a distance of nearly 900 light years.

Magnitude 3.79 **Misam (κ Persei)** is a yellow giant star shining from a distance of 113 light years.

Delta (δ) Persei is a blue giant star shining at magnitude 3.01 from a distance of 512 light years.

Another blue giant is **Epsilon (ε) Persei**, the magnitude 2.90 glow of which set off towards us nearly 640 years ago.

minima of Algol do not often take place at convenient times, with many occurring during daylight hours and only a few actually taking place during the evening. Predictions of the minima of Algol are available in numerous publications and it may be better to arm yourself with these before planning to check out this most famous of variable stars!

OPEN STAR CLUSTER M34

The open star cluster Messier 34 (M34) or NGC 1039 lies a little to the west of the star Algol, shining with an overall magnitude of 5.5 from a distance of around 1,500 light years. When he discovered this object in 1764, Charles Messier described it as: *'A cluster of small stars . . . '.*

If the sky is really dark and clear M34 may just be visible without any optical aid, as attested by the German astronomer Johann Elert Bode who described this

object as: *'A star cluster, visible to the naked eye'.* However, a pair of binoculars, together with the finder chart, will undoubtedly make your search easier. Start by identifying the guide stars Algol, Misam and Pi (π) Persei and, once these have been located, use the chart to help you star hop your way to the cluster. Binoculars should show M34 as a distinct nebulous patch and even a small telescope will resolve some of its individual member stars.

THE SWORD HANDLE DOUBLE CLUSTER

The two open star clusters NGC 869 and NGC 884 are visible to the naked eye under really clear, dark skies and can be found near the border between Perseus and the neighboring constellation Cassiopeia. Records of these two objects go as far back as the second century AD

when the Greek astronomer Ptolemy referred to them as being: *'At the tip of the right hand (of Perseus) and is misty [nebulosa]'.*

Both NGC 869 and NGC 884 are located at a distance of around 7,500 light years, their overall magnitudes being 4.4 (NGC 869) and 4.7 (NGC 884). Binoculars will show the clusters quite well and small telescopes will reveal much of the splendor of these two beautiful objects. Although they lie quite close to each other in the sky, a wide field of view is needed in order to see both clusters at once. Known collectively as the Sword Handle Double Cluster, and depicting the hand of Perseus as he holds his sword aloft, they lie in a rich section of the Milky Way and time spent sweeping this area with binoculars, or a telescope at low magnification, will be well rewarded.

OPEN STAR CLUSTER NGC 1528

The open star cluster NGC 1528 is a fairly easy target for the backyard astronomer. Located near the northeastern border of Perseus, and shining at magnitude 6.4, it can be found a little to the north of the two stars Mu (μ) and Lambda (λ) Persei. Once you have identified these two stars, both of which are shown on the finder chart *(see page 170)*, you can star hop your way to NGC 1528 through the field of fainter stars in the immediate vicinity. Binoculars will reveal the cluster as a dim and slightly oval patch of light although even a small telescope will resolve some of its individual member stars.

OPPOSITE LEFT: Open star cluster M34 (NGC 1039) in Perseus.

OPPOSITE RIGHT: The magnificent Sword Handle Double Cluster NGC 869 and NGC 884.

BELOW: Located at a distance of around 1,500 light years, the California Nebula (NGC 1499) is an emission nebula in Perseus.

PHOENIX
The Phoenix

Representing the legendary bird of the same name, the constellation Phoenix lies slightly to the north of a line between Fomalhaut in Piscis Austrinus and the bright star Achernar in Eridanus, both stars being depicted on the chart to help you to locate the group. According to mythology, the Phoenix lived a long life, following which it built a nest of spices. It then set fire to the nest, during which process the bird was burned to ashes and came forth again with new life.

Phoenix is another of the groups introduced by Pieter Dirkszoon Keyser and Frederick de Houtman following their expedition to the East Indies in the 1590s. Both Fomalhaut and Achernar, as well as Phoenix itself, can be seen from anywhere south of latitude 32°N and the entire constellation can be seen from most of Mexico, northern Africa and India and from locations south of these. Phoenix is always low in the sky when viewed from places north of the equator, the best views of this faint group being from regions much further south than this.

PHOENIX

THE STARS OF PHOENIX

Its leading star is **Ankaa (α Phoenicis)**, the name of which is derived from the Arabic **'al-anqa'** meaning 'the Fabulous Bird', an allusion to the constellation as a whole. This orange giant star shines at magnitude 2.40 from a distance of 85 light years.

Magnitude 3.32 yellow giant **Beta (β) Phoenicis** is the second brightest star in the group, slightly to the northeast of which is the red giant **Gamma (γ) Phoenicis**, the magnitude glow of which has taken around 230 years to reach us.

Epsilon (ε) Phoenicis is an orange giant shining at magnitude 3.88 from a distance of around 145 light years.

Shining at magnitude 3.93, the light from yellow giant **Delta (δ) Phoenicis** reaches us from a distance of a little over 140 light years.

Magnitude 4.36 **Eta (η) Phoenicis** lies at a distance of 245 light years.

AN ALGOL-TYPE VARIABLE

Phoenix plays host to the Algol-type (see entry for Perseus) variable **Zeta (ζ) Phoenicis**, which can be found just to the northwest of the bright star Achernar and marking the southern corner of the main diamond-shape of Phoenix. Located at a distance of nearly 300 light years, this eclipsing binary ranges in magnitude from 3.92 to 4.42 over a period of 1.67 days (a little over 40 hours). The eclipse phase is easily visible to the naked eye and comparisons of the changing brightness of Zeta can be made with the nearby Delta (magnitude 3.93) and Eta (magnitude 4.36) as well as Epsilon (magnitude 3.88), located some way to the northwest of these.

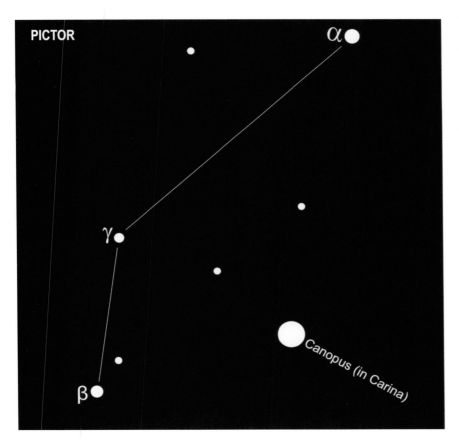

THE STARS OF PICTOR

Alpha (α) Pictoris is the brightest star in the group, shining at magnitude 3.24 from a distance of 96 light years.

Somewhat closer to us is **Beta (β) Pictoris**, a magnitude 3.85 star whose light has taken just 63 years to reach us.

The magnitude 4.50 orange giant **Gamma (γ) Pictoris** completes the main trio of stars in Pictor, the light from this star having set off towards us around 180 years ago.

PICTOR
The Painter's Easel

The tiny constellation Pictor is another of the constellations devised by the French astronomer Nicolas Louis de Lacaille and is intended to represent a painter's easel and palette. Located just to the west of the star Canopus in Carina, this group is unimaginative to say the least. Canopus is included on the chart as a guide to locating Pictor.

Taking the form of a bent line of three faint stars, Pictor is visible in its entirety from locations to the south of latitude 26°N, making it accessible to backyard astronomers in the equatorial regions and south from there. Observers in South America, South Africa, Australia and New Zealand will see the brilliant star Canopus high in the sky just to the south of the overhead point during late evenings in December. Once Canopus is located the unimpressive collection of stars forming Pictor can be found.

PISCES
The Fishes

Pisces is the 14th largest of the constellations and can be found trailing around the southeastern borders of the neighboring constellation Pegasus. The two stars 82 and Tau (τ) Piscium, located just to the south of the bright star Mirach in Andromeda, represent one fish, the other depicted by a circlet of stars lying immediately to the south of the Square of Pegasus. Old star charts show the two fish tied together by a cord, although any significance that may have once been placed in this cord is not known. Pisces lies just to the north of the celestial equator and is completely visible from almost all inhabited areas of the world. The chart shows Pisces, together with the Square of Pegasus and the star Mirach in Andromeda, which between them should enable you to locate the winding form of this faint group.

Legend associates Pisces with Aphrodite and her son Eros who were caught up in the battle between Zeus and the fearsome monster Typhon, this being the battle during which Pan assisted Zeus *(see Capricornus)*. During the battle the pair took refuge from Typhon by hiding in reeds bordering the river, although it is here that the legend becomes uncertain. Some accounts say that Aphrodite and Eros turned into fishes and swam to safety, others stating that the pair were carried to safety by two fishes who were swimming by at the time. Whatever the source or identity of the two fishes, they are represented by the constellation Pisces.

DOUBLE STARS IN PISCES

Psi[1] (ψ) Piscium has two 5th magnitude components which are easily resolved in a small telescope and may be glimpsed in binoculars. Another easy double for small telescopes is **Zeta (ζ) Piscium** which has 5th and 6th magnitude components. **Kappa (κ) Piscium**, the Circlet's southernmost star, shines at magnitude 4.95 from a distance of around 160 light years. Binoculars or a small telescope will reveal a nearby companion star, although the proximity of this star to Kappa is nothing more than a line of sight effect. The magnitude 6.26 orange giant component lies at a distance in excess of 350 light years, putting it over twice as far away as Kappa itself.

THE STARS OF PISCES

The yellow giant **Al'farg (η Piscium)** is the brightest star in Pisces, its magnitude 3.62 glow reaching us from a distance of almost 350 light years.

Slightly fainter is magnitude 3.70 **Gamma (γ) Piscium**, the light from which has taken around 140 years to reach us.

Deriving its name from the Arabic *'al-risha'* meaning 'the Cord', **Alrescha (α Piscium)** shines at magnitude 3.82 from a distance of 150 light years. The name of this star reflects its location within the constellation, depicting as it does the point at which two cords joining the fishes meet.

The 'Circlet' is the most prominent part of Pisces and represents the head of the western fish. Here we find the magnitude 4.48 star **Fum al Samakah (β Piscium)**. Located at a distance of over 400 light years, this star takes its name from the Arabic for 'the Fish's Mouth' which is appropriate bearing in mind its position relative to the westernmost of the two fishes.

The imaginatively-named **Torcularis Septentrionalis (o Piscium)** is a magnitude 4.26 yellow giant whose light set off towards us around 280 years ago. Its name is Latin for 'the Northern Press' although the reasons behind the star being given this strange name are lost in antiquity.

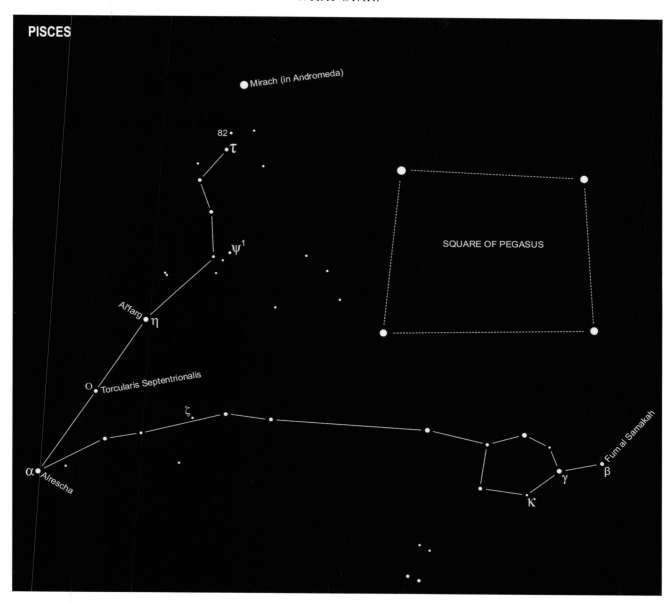

PISCES

Mirach (in Andromeda)

82

τ

ψ¹

Al'farg η

O Torcularis Septentrionalis

ζ

α Alrescha

κ

γ

β Fum al Samakah

SQUARE OF PEGASUS

PISCIS AUSTRINUS
The Southern Fish

The constellation Piscis Austrinus can be viewed in its entirety from anywhere south of latitude 53°N, and those observing from Argentina, South Africa, Australia and similar latitudes during October evenings will see brilliant Fomalhaut at or near the overhead point. However, backyard astronomers at mid-northern latitudes will see Piscis Austrinus lying low over the southern horizon, and managing to locate it at all may present something of a challenge unless the sky is particularly dark and clear.

OPPOSITE: One of the most prominent stars in the sky, and depicting the mouth of the Southern Fish, Fomalhaut is the brightest star in the constellation Piscis Austrinus.

THE STARS OF PISCIS AUSTRINUS

Fomalhaut (α Piscis Austrini), the brightest star in the constellation Piscis Austrinus can be located by following a line drawn from Scheat through Markab, both in the Square of Pegasus. The line passes immediately to the east of the 3rd magnitude star Skat in Aquarius, as shown here, before reaching Fomalhaut. Old star charts depict the Southern Fish lying on its back drinking the liquid being poured into its mouth by Aquarius. This is reflected in the name Fomalhaut, which is derived from the Arabic **'fam al-hut al-janubi'** meaning 'the Mouth of the Southern Fish'. Fomalhaut itself shines at magnitude 1.17 from a distance of 25 light years.

The next brightest star in the constellation is **Epsilon (ε) Piscis Austrini**. Shining at magnitude 4.18 and located just to the northwest of Fomalhaut, the light from this star has taken nearly 500 years to reach our planet.

The yellow giant star **Delta (δ) Piscis Austrini** shines at magnitude 4.20 from a distance of around 150 light years.

Iota (ι) Piscis Austrini, a white giant star shining at magnitude 4.35 from a distance of just over 200 light years, marks the westernmost point of the constellation.

Although Piscis Austrinus offers little to the backyard astronomer other than identification of the constellation itself, the double star **Beta (β) Piscis Austrini** is worth a closer look, its magnitude 4.4 and 7.9 pale yellow and white components being resolvable in small telescopes.

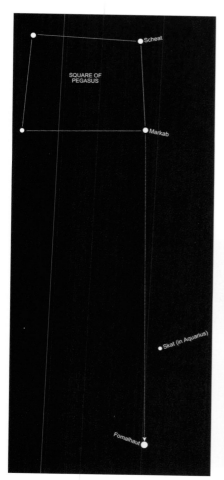

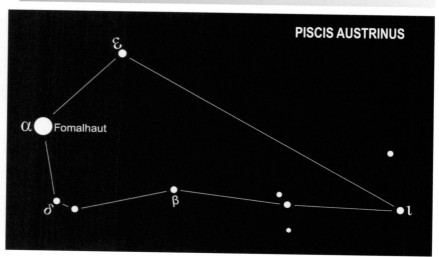

(Part of) CANIS MAJOR

Muliphein

Sirius

Mirzam

4

M46

M47

ξ

Azmidiske

M93

k^{1,2}

Markeb

Naos ζ

ν

ι²

ι¹

τ

Canopus (in Carina)

PUPPIS

PUPPIS
The Poop or Stern

Puppis is bordered by Carina to the south and Vela to the southeast and is the largest of the three constellations created when Lacaille dismantled Argo Navis *(see Carina)*. The constellation extends from a point near Canopus in Carina, along the Carina/Vela borders and northwards past

THE STARS OF PUPPIS

Deriving its name from the Greek for 'Ship', **Naos (ζ Puppis)** is the brightest star in Puppis, this magnitude 2.21 blue supergiant star shining from a distance of over 1,000 light years.

The magnitude 3.34 glow of the yellow supergiant **Azmidiske (ξ Puppis)** has taken over 1,100 years to reach us.

Located just to the southwest of Azmidiske is **Markeb (k^{1,2} Puppis)**, a double star with components of magnitudes 4.5 and 4.6 which can be resolved in a small telescope. A number of the stars in Puppis have English designations, that of k Puppis occasionally (and erroneously) being written as Kappa (κ) rather than 'k'.

Tau (τ) Puppis is an orange giant shining at magnitude 2.94 from a distance of 182 light years.

The light from magnitude 3.17 blue giant **Nu (ν) Puppis** set off towards us around 370 years ago.

Sirius in Canis Major. Puppis can be seen in its entirety from latitudes south of 39°N, putting it within visual reach of observers in the central United States, southern Europe and locations to the south of these. Although Puppis is generally regarded as a southern group, its northern extremities can be seen from mid-northern latitudes a little way to the east of the trio of stars Sirius, Mirzam and Muliphein in Canis Major. These three stars are included on the chart, along with the bright star Canopus in Carina, as guides to locating Puppis.

OPEN STAR CLUSTERS M46 & M47

The two open star clusters **Messier 46** (M46) or NGC 2437 and **Messier 47** (M47) or NGC 2422 can be seen to the east of Sirius in the neighboring constellation Canis Major, near the northern border of Puppis. Although the discoveries of both M46 and M47 are credited to Charles Messier in 1771, M47 was originally discovered by the Italian astronomer Giovanni Batista Hodierna during the early 17th century.

M47 shines at 4th magnitude from a distance of around 1,600 light years, and can be glimpsed with the naked eye if the sky is really dark and clear. Somewhat fainter is M46 although, with an overall magnitude of around 6, this object is an easy target for binoculars or a small telescope. M46 lies at a distance of over 5,000 light years and contains up to around 500 stars, making it somewhat

ABOVE: Open star cluster M47 (NGC 2422).

RIGHT: Open star cluster M93 (NGC 2447).

larger than M47 which has around 50 member stars.

Extending a line eastwards from Mirzam, through Sirius and Muliphein in Canis Major and almost as far again, will lead you to the small triangle of faint stars which includes 4 Puppis. M46 and M47 lie within this triangle and can be seen together in the wide field of view of a pair of binoculars.

OPEN STAR CLUSTER M93

Located at a distance of around 3,600 light years and containing around 50 individual stars, the open star cluster **Messier 93** (M93) or NGC 2447 lies just to the northwest of the star Azmidiske. This cluster was discovered in 1782 by Charles Messier who described it as: '*A cluster of small stars . . . between Canis Major and the*

prow of Navis'. With an overall magnitude of around 6, M93 is just about visible to the naked eye, although it would be quite difficult to see without optical aid unless the sky was very dark and clear. To locate M93 in binoculars, look carefully at the region of sky adjoining the star Azmidiske where M93 should reveal itself as a misty patch of light.

THE SEMI-REGULAR VARIABLE STAR L² PUPPIS

Lying immediately to the north of the magnitude 4.87 star L^1, and forming a small triangle with the two stars Nu and Tau at the southern end of Puppis (a little to the northeast of Canopus in Carina), is the semi-regular variable star L^2 Puppis, the orange color of which is evident when viewed through binoculars. Shining from a distance of just over 200 light years, L^2 varies between 3rd and 6th magnitudes over a period of a little over 140 days, its complete cycle of variability observable either with the naked eye or through binoculars.

PYXIS
The Mariner's Compass

(See Antlia)

RETICULUM
The Net

The tiny constellation Reticulum is yet another of the groups introduced into this region of sky by the French astronomer Nicolas Louis de Lacaille during his stay in South Africa. Taking the form of an approximately diamond-shaped pattern of stars, it represents the reticule used in the eyepiece of his telescope by Lacaille to measure the positions of stars.

Reticulum borders the neighboring constellation Hydrus, the main stars of which, together with the brilliant Achernar in Eridanus, are included on the chart to help you identify the group. The whole of the constellation Reticulum, together with the guide stars depicted here, are visible to observers within or south of the Earth's equatorial regions.

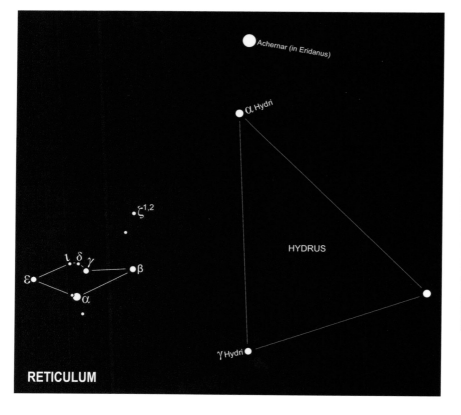

RETICULUM

THE STARS OF RETICULUM
The brightest star in Reticulum is **Alpha (α) Reticuli**, a magnitude 3.33 yellow giant shining from a distance of around 160 light years.

Beta (β) Reticuli is an orange giant star, the magnitude 3.84 glow of which has taken 100 years to reach us

Magnitude 4.44 orange giant **Epsilon (ε) Reticuli** lies at a distance of just 59 light years.

Gamma (γ) Reticuli is a magnitude 4.48 red giant, the light from which has taken around 470 years to reach us.

Another red giant is **Delta (δ) Reticuli** which shines at magnitude 4.56 from a distance of just over 500 light years.

Rounding off the main form of Reticulum is **Iota (ι) Reticuli**, a magnitude 4.97 orange giant located at a distance of a little over 300 light years.

Lying at a distance of 39 light years, **Zeta^{1,2} (ζ^{1,2}) Reticuli** is a wide binary with yellow components of magnitudes 5.24 and 5.53 which are easily split with binoculars. The two stars in the Zeta system are just about far enough apart to be resolvable with the naked eye under really dark, clear skies.

SAGITTA
The Arrow

Although quite small and faint, Sagitta has a distinctive shape. Resembling a tiny dart, it is one of the few constellations that actually resembles the object that it is supposed to depict. Sagitta is bordered to the south by the constellation Aquila, the

THE STARS OF SAGITTA

Gamma (γ) Sagittae is the brightest star in Sagitta, the light from this magnitude 3.51 red giant reaching us from a distance of 258 light years.

Shining with a magnitude of 3.68, **Delta (δ) Sagittae** lies at a distance of 580 light years.

Sham (α Sagittae) and **Beta (β) Sagittae** both shine at magnitude 4.39 and lie at similar distances from us. Deriving its name from the Arabic *'al-sahm'* meaning 'the Arrow', the light from yellow giant Sham set off on its journey towards us 425 years ago, putting it slightly closer than Beta which shines from a distance of around 440 light years.

Zeta (ζ) Sagittae is a magnitude 5.01 white star located at a distance of around 250 light years.

Epsilon (ε) Sagittae is a wide optical double star which is easily resolved in binoculars. Situated near the southwestern border of Sagitta, and at a distance of around 470 light years, the primary component of Epsilon is a magnitude 5.70 yellow giant, which is accompanied by an 8th magnitude bluish-white companion.

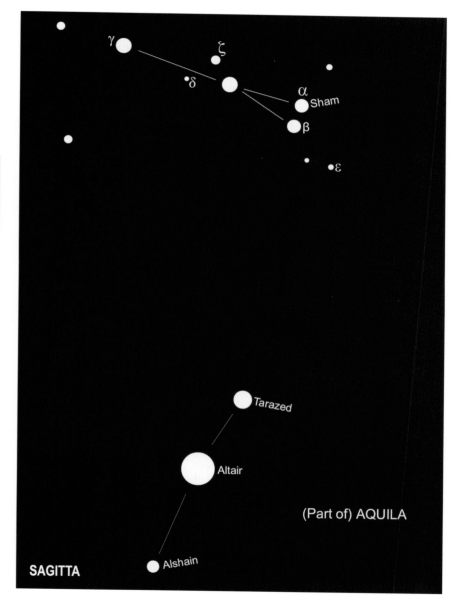

183

trio of bright stars Alshain, Altair and Tarazed in Aquila depicted here as a guide. Because Sagitta is located only a little way to the north of the celestial equator, the whole of this tiny constellation is visible from any inhabited part of the world.

One of the legends attached to Sagitta identifies it as one of the arrows with which Hercules slew the Stymphalian Birds. Stymphalus was a lake upon the banks of which lived a flock of dangerous brazen-clawed birds. Hercules wiped out the entire flock using arrows

that had been dipped in the venomous blood of Hydra, whom Hercules had slain as his second labor.

BELOW: Globular cluster M71 (NGC 6838) in Sagitta.

SAGITTARIUS
The Archer

Situated a little to the south of the celestial equator, and bordered to the west by the constellation Scorpius, Sagittarius can be seen fairly low down in the sky from mid-northern latitudes. The group is visible in its entirety from locations south of latitude 45°N although it is best viewed from the southern hemisphere from where it can be seen high up in the northern sky during the winter months. Parts of Sagittarius are visible from almost every inhabited part of the world, several stars in the 'sting' of the neighboring Scorpius, including Shaula and Sargas, being shown here as a guide to locating the group. Sagittarius was one of the 48 constellations drawn up by the Greek astronomer Ptolemy during the second century. Legend associates this group with an archer firing an arrow in the direction of Scorpius, the archer in question being the satyr Crotus, the son of Pan. Crotus was an accomplished hunter and the inventor of archery, his prowess in hunting and skill with a bow and arrow eventually leading to him being placed in the sky by Zeus.

THE TEAPOT

Although Sagittarius is a fairly large and rambling constellation, with no clear and well-defined shape, its central region does contain one of the most distinctive and well-known asterisms. The asterism in question is the Teapot, the four stars Ascella, Phi (φ) Sagittarii, Kaus Media and Kaus Australis making up the body of the Teapot, the top of the lid represented by Kaus Borealis, the handle by Tau (τ) and Sigma (σ) Sagittarii and the tip of the spout by Nash. The eight stars forming the Teapot stand out quite well and act as a good guide to identifying the constellation as a whole.

THE TRIFID NEBULA

Sagittarius plays host to more Messier objects than any other constellation, one of these being the gaseous nebula **Messier 20** (M20) or NGC 6514. Also known as the Trifid Nebula, M20 is located a little way to the southwest of the star Mu (μ) Sagittarii and appears to have been originally discovered by the French astronomer Guillaume-Joséph-Hyacinthe-Jean-Baptiste Le Gentil de la Galaisière prior to 1750, although was independently found by Charles Messier who added it to his catalogue in 1764. M20 lies at a distance of around 5,000 light years and, with an overall magnitude of 6.3, is within the light grasp of binoculars or a small telescope.

The main section of M20 consists of a characteristically colored red area of emission nebulosity (see Glossary) which,

THE STARS OF SAGITTARIUS

Deriving their names from a mixture of Arabic and Latin meaning the southern, middle and northern part(s) of the bow, the three stars Kaus Australis, Kaus Media and Kaus Borealis together represent the bow held by Sagittarius. Magnitude 1.79 white giant **Kaus Australis (ε Sagittarii)** is the brightest star in Sagittarius, its light reaching us from a distance of 143 light years. The orange giant **Kaus Media (δ Sagittarii)** lies immediately to the north and shines at magnitude 2.72 from a distance of around 350 light years. The light from magnitude 2.82 **Kaus Borealis (λ Sagittarii)**, located slightly further to the north, set off towards us 78 years ago.

Positioned at the end of the arrow being fired by the archer and taking its name from the Arabic for 'the Point' is the star **Nash (γ Sagittarii)**, a magnitude 2.98 orange giant shining from a distance of 97 light years.

Binoculars will resolve the pretty double formed from the two stars **Arkab Prior (β¹ Sagittarii)** and **Arkab Posterior (β² Sagittarii)**. Taking their names from the Arabic for 'the Tendon', these two stars can be found near the southern border of Sagittarius where they represent the archer's tendons. Arkab Prior is itself a binary star, its magnitude 3.96 and 7.40 components resolvable in a small telescope. Arkab Posterior shines at magnitude 4.27 and lies at a distance of 134 light years, roughly a third of the distance of magnitude Arkab Prior, the light from which has taken around 375 years to reach us.

Magnitude 3.96 **Rukbat (α Sagittarii)** derives its name from the Arabic *'rukbat al-rami'* meaning 'the Knee of the Archer'. Rukbat lies at a distance of around 180 light years, putting it roughly twice as far away as magnitude 2.60 **Ascella (ζ Sagittarii)**, the light from which has taken just 88 years to reach us. Ascella takes its name from the Latin for 'the Armpit'.

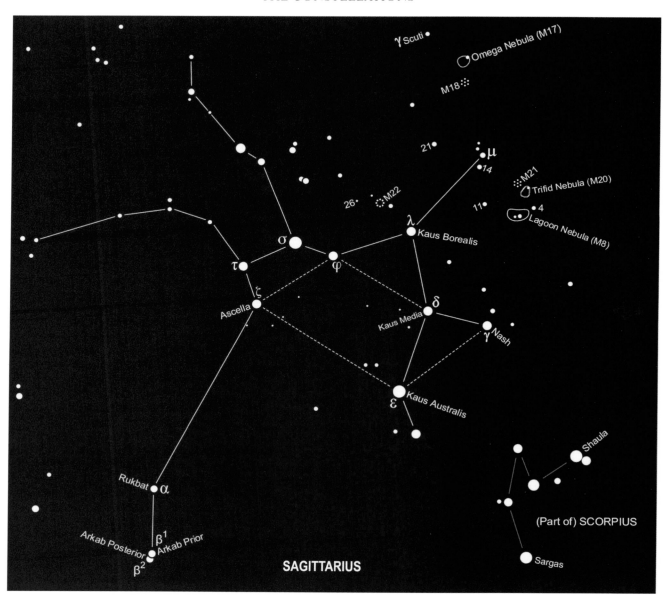

γ Scuti

Omega Nebula (M17)

M18

21

μ

14

26

M22

λ Kaus Borealis

M21

Trifid Nebula (M20)

11

4

Lagoon Nebula (M8)

σ

τ

φ

ζ

Ascella

δ

Kaus Media

γ Nash

ε Kaus Australis

Shaula

Rukbat α

(Part of) SCORPIUS

β¹

Arkab Posterior

Arkab Prior

β²

Sargas

SAGITTARIUS

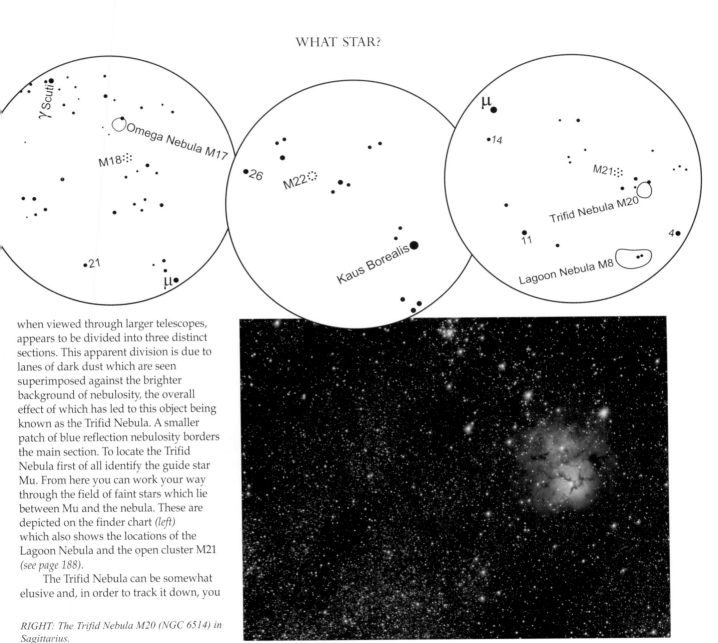

when viewed through larger telescopes, appears to be divided into three distinct sections. This apparent division is due to lanes of dark dust which are seen superimposed against the brighter background of nebulosity, the overall effect of which has led to this object being known as the Trifid Nebula. A smaller patch of blue reflection nebulosity borders the main section. To locate the Trifid Nebula first of all identify the guide star Mu. From here you can work your way through the field of faint stars which lie between Mu and the nebula. These are depicted on the finder chart (left) which also shows the locations of the Lagoon Nebula and the open cluster M21 (see page 188).

The Trifid Nebula can be somewhat elusive and, in order to track it down, you

RIGHT: The Trifid Nebula M20 (NGC 6514) in Sagittarius.

will need a clear, dark and moonless sky. As is generally the case with objects of this type, you will need to look for a patch of light rather than a star-like point. It should also be pointed out that the different colors of the nebulosity may not be particularly evident when viewed through binoculars or small telescopes.

OPEN STAR CLUSTER M21

Discovered by Charles Messier in 1764, and situated immediately to the northeast of the Trifid Nebula, is the open star cluster **Messier 21** (M21) or NGC 6531. Located at a distance of over 4,000 light years and containing over 50 member stars, M21 shines with an overall magnitude of around 7 and is reasonably easy to locate with binoculars or a small telescope.

THE LAGOON NEBULA

Discovered by the Sicilian astronomer Giovanni Battista Hodierna prior to 1654, the gaseous emission nebula **Messier 8** (M8) or NGC 6523 was added to the catalogue of Charles Messier in 1764. Also known as the Lagoon Nebula, this object shines at around 6th magnitude and can be glimpsed with the naked eye under really dark, clear skies. Binoculars reveal M8 as a roughly oval, colorless patch of light surrounding a bright star cluster. Under favorable conditions, even a small telescope will reveal a dark dust lane separating this region from a second area of nebulosity.

The Lagoon Nebula has the reddish color which is characteristic of objects of this type, although long-exposure photographs are needed in order to bring

this out. Situated just to the south of M20, and located at a distance of around 5,000 light years, the Lagoon Nebula is regarded as one of the finest nebulae in the sky and can be found by carefully following the curve of faint stars extending away to the southwest of Mu Sagittarii. The Lagoon Nebula forms a tiny triangle with the nearby Trifid Nebula and the faint star 4 Sagittarii.

THE OMEGA NEBULA

Discovered by Philippe Loys de Cheseaux in 1746, and located near the border between Sagittarius and the neighbouring constellation Scutum, is the emission nebula **Messier 17** (M17) or NGC 6618.

OPPOSITE: The Lagoon Nebula M8 (NGC 6523).

ABOVE: The Omega Nebula M17 (NGC 6618).

When Charles Messier added it to his catalogue in 1764 he described this object as: *'A train of light without stars . . . in the shape of a spindle . . . light very faint'.*

M17 is also known as the Omega Nebula or Horse-shoe Nebula and, shining at around 6th magnitude, it is just about visible to the naked eye provided the sky is exceptionally dark and clear. The Omega Nebula is smaller and more concentrated than the Lagoon Nebula and is consequently that little bit easier to see.

It can be found either by using the star Gamma (γ) Scuti in the neighboring constellation Scutum as a guide, or by locating Mu (μ) Sagittarii from where you can star hop your way northwards towards the nebula. Both guide stars are depicted on the finder chart *(page 187)*, which also shows the open star cluster M18. Although not revealing the characteristic reddish color of the Omega Nebula, binoculars or a small telescope will show this object reasonably well.

OPEN STAR CLUSTER M18

The open star cluster **Messier 18** (M18) or NGC 6613 lies just to the south of the Omega Nebula. Discovered by Charles Messier in 1764 and described by him as: '*A cluster of small stars, a little below M17 . . .*', M18 shines at magnitude 7.5 from a distance of 4,900 light years. Although M18 is not very large and only contains around 20 stars or so, it is seen superimposed against the backdrop of the Milky Way. Such is the beauty of this region that the English astronomer Thomas William Webb, when alluding to the cluster, described the surrounding area as a: '*Glorious field in a very rich vicinity*'. As is the case with the Omega Nebula, M18 can be found by working your way either northwards from Mu Sagittarii or southwards from Gamma Scuti. Binoculars will enable you to track this cluster down and a small telescope will resolve a dozen or so of its member stars.

GLOBULAR CLUSTER M22

Sagittarius is home to a number of globular star clusters, one of which is **Messier 22** (M22) or NGC 6656 which can be found a little to the east of the star Kaus Borealis. Discovered by the German amateur astronomer Abraham Ihle in 1665, it may have been seen earlier by the Polish astronomer Johannes Hevelius, Messier adding it to his catalogue in 1764. Containing 70,000 or more individual stars, M22 shines at magnitude 5.9 from a distance of around 10,500 light years and appears to be the first ever globular cluster to be discovered.

M22 is an fairly easy target for binoculars, which will show it as a

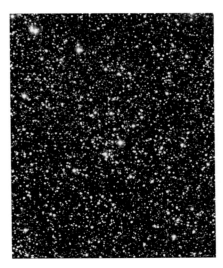

prominent but fuzzy patch of light, and you may even catch a glimpse of it with the naked eye providing the sky is really dark, clear and moonless. When viewed through small telescopes a number of its individual member stars will be revealed. Use the finder chart (*page 187*) to guide you to the cluster, working your way to the northeast from the guide star Kaus Borealis and towards the faint star 26 Sagittarii.

OPPOSITE: A Hubble Space Telescope image capturing a small region within the Omega Nebula.

LEFT: Open star cluster M18 (NGC 6613).

BELOW: Globular cluster M22 (NGC 6656).

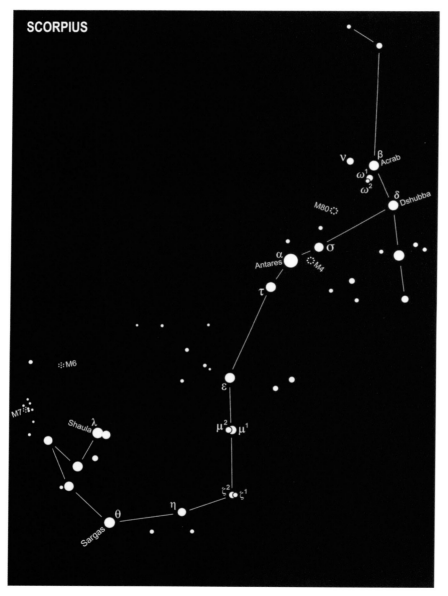

SCORPIUS
The Scorpion

Scorpius represents the scorpion commanded by Diana, the Goddess of the Hunt, to sting and kill Orion following their hunting match *(see Orion)*, the overall layout of Scorpius indeed resembling that of a scorpion. Scorpius is visible in its entirety from latitudes south of 44°N and can be viewed from the central United States, southern Europe and most of Japan and from latitudes to the south of these areas.

Scorpius was originally much larger and included a number of stars which depicted the scorpion's claws *(see Libra)*. Known as Chelae, meaning 'The Claws of the Scorpion', these stars were removed from Scorpius by the Romans who used them to form the separate constellation Libra which can be found bordering Scorpius to the west.

ABOVE: A Hubble Space Telescope image of globular cluster M4 (NGC 6121) in Scorpius.

THE STARS OF SCORPIUS

Scorpius is not a particularly large constellation although it is very prominent, especially when viewed from the equatorial regions or from locations further to the south. The brightest star in the constellation is **Antares (α Scorpii)**, a red supergiant shining with a magnitude of around 1.06 from a distance of 535 light years. The name of this star can be loosely translated as 'Rival of Mars', indicative of the fact that, due to its strong red color, it vies for prominence with Mars (also known as the Red Planet) when the two are seen in the same area of sky.

Deriving its name from the Arabic *'al-shaula'* meaning 'the Scorpion's Sting', magnitude 1.62 **Shaula (λ Scorpii)** shines from a distance of around 550 light years.

Dschubba (δ Scorpii) takes its name from the Arabic *'jabhat al-aqrab'* meaning 'the Scorpion's Forehead', the light from this magnitude 2.29 star having set off towards us almost 400 years ago.

Located just to the northwest of Antares is **Sigma (σ) Scorpii** which shines at magnitude 2.90 from a distance of just under 700 light years. Somewhat closer is magnitude 2.82 **Tau (τ) Scorpii**, located just to the southeast of Antares and the light from which has taken only around 450 years to reach us. Sigma and Tau share the name 'Alniyat', taken from the Arabic *'al-niyat'* meaning 'Outworks of the Heart' and protect Antares, regarded as being the heart of Scorpius. Sigma, Antares and Tau form a conspicuous trio of stars.

Epsilon (ε) Scorpii is an orange giant star, the magnitude 2.29 glow of which set off towards us 64 years ago.

The light from magnitude 3.32 **Eta (η) Scorpii** reaches us from a distance of 73 light years.

Sargas (θ Scorpii) is a magnitude 1.86 white giant shining from a distance of 270 light years.

DOUBLE STARS IN SCORPIUS

Acrab (β Scorpii) is a double star with components of magnitudes 2.6 and 4.9 which can be resolved in small telescopes. Located at a distance of around 400 light years, the name of this star is derived from the Arabic *'al-aqrab'* meaning 'the Scorpion'.

Located just to the east of Acrab is **Nu (ν) Scorpii**, a double star with magnitude 6.5 and 4.4 components shining from a distance of around 450 light years. Both components of Nu are easily resolved in binoculars, although closer examination with a large telescope will reveal that each of the two stars forming Nu is actually double again, the whole system resembling the famous 'double-double' star Epsilon in the constellation Lyra.

Mu (μ) Scorpii is a wide double star which can be resolved with the naked eye. Located immediately to the south of Epsilon, its brightest component is magnitude 3.0 **Mu[1]** which is a little more prominent than its magnitude 3.6 companion **Mu[2]**. The apparent proximity to each other of these two stars is nothing more than a line of sight effect, Mu[2] having been found to lie around 15 light years further away from us.

Located near the southern border of Scorpius, **Zeta (ζ) Scorpi** has a brighter magnitude 3.6 orange component (Zeta[2]) and a fainter magnitude 4.7 blue-white companion (Zeta[1]) both of which can be resolved with the naked eye. Again these two stars are not physically related, **Zeta[1]** shining from a distance of around 150 light years, putting it considerably closer to us than Zeta[2], the light from which set off towards us around 5,000 years ago.

Another naked-eye double, and an excellent target for binoculars, is that formed from the two stars **Omega[1] (ω[1])** and **Omega[2] (ω[2]) Scorpii**, which lie just to the south of Acrab. Omega[1] is a blue magnitude 3.93 star shining from a distance of around 470 light years. A little closer to us is the yellowish magnitude 4.31 Omega[2], the light from which has taken 290 years to reach us.

GLOBULAR CLUSTER M4

Discovered in around 1746 by the Swiss astronomer Jean-Philippe Loys de Cheseaux, the globular cluster **Messier 4 (M4)** or NGC 6121 lies immediately to the west of Antares. Charles Messier added it to his catalogue in 1764, describing it as a *'Cluster of very small stars; with an inferior telescope it appears more like a nebula . . . '* With a diameter of around 75 light years, M4 shines from a distance of a little over 7,000 light years. Its proximity to Antares makes it very easy to locate in binoculars which will reveal this 6th magnitude object as a fuzzy ball of light, even a small telescope bringing out some of its individual member stars.

THE PTOLEMY CLUSTER

The open star cluster **Messier 7** (M7) or NGC 6475 was first recorded by the Greek astronomer Ptolemy during the second century AD, which has led to it being sometimes known as the Ptolemy Cluster. Ptolemy described this magnitude 3.3 object as being ' . . . *misty* . . . ' and ' . . . *following the Sting (of Scorpius) . . .* '. Containing around 80 stars and located at a distance of around 800 light years, M7 is easily visible to the naked eye as a hazy patch of light a little way to the northeast of the bright star Shaula in the Sting of Scorpius.

THE BUTTERFLY CLUSTER

Visible to the naked eye a little to the northwest of M7 is the magnitude 4.2 open star cluster **Messier 6** (M6) or NGC 6405. Shining from a distance of around 1,600 light years, M6 forms a triangle with M7 and the nearby star Shaula and, together with M7, may have once formed the termination of the sting of the Scorpion. Also known as the Butterfly Cluster, due to its overall shape vaguely resembling that of a butterfly, this object was recorded by the Italian astronomer Giovanni Batista Hodierna during the early part of the 17th century and again by the Swiss astronomer Philippe Loys de Cheseaux in around 1746, Charles Messier adding both M6 and M7 to his catalogue in 1764.

OPPOSITE: Open star cluster M7 (NGC 6475).

ABOVE RIGHT: Globular cluster M80 (NGC 6093).

RIGHT: Open star cluster M6 (NGC 6405).

Many of the 100 or so stars in the M6 cluster can be resolved in binoculars although, as far as observers at mid-northern latitudes are concerned, both M6 and M7 are seen fairly low down over the southern horizon and may be difficult to spot if there is any mist, cloud or light pollution present. They are best viewed from the southern hemisphere from where these two fine clusters are well placed, high in the sky and away from any potential horizon glow.

GLOBULAR CLUSTER M80

Forming a triangle with the stars Dschubba and Antares, and lying roughly equidistant from the two, is the magnitude 7.3 globular cluster **Messier 80** (M80) or NGC 6093. Discovered by Charles Messier in 1781, M80 shines from a distance of over 32,000 light years and can be tracked down fairly easily in binoculars providing the sky is reasonably dark and clear. When seen through binoculars or a small telescope, M80 takes on the appearance of a hairy star, Messier himself describing it as a *'Nebula without star in the Scorpion . . .'*. Take a look at M80 for yourself and see if your impression of this object matches that of Messier.

SCULPTOR
The Sculptor

This is one of the constellations devised by the French astronomer Nicolas Louis de Lacaille during the 1750s and was intended to represent a sculptor's workshop. This group was originally called Apparatus Sculptoris, a name which has since been shortened to Sculptor. It takes the form of a curved line of faint stars, none of which are named and none of which are prominent.

Sculptor is situated immediately to the south of the star Deneb Kaitos in

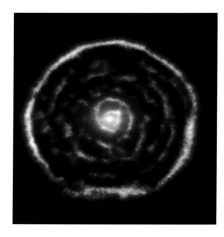

THE STARS OF SCULPTOR

Shining at magnitude 4.30 is the blue giant **Alpha (α) Sculptoris**, the brightest star in Sculptor, the light from which reaches us from a distance of over 750 light years.

The light from magnitude 4.38 **Beta (β) Sculptoris** has taken 175 years to reach us.

Gamma (γ) Sculptoris is an orange giant, its magnitude 4.41 glow emanating from a distance of 180 light years.

Delta (δ) Sculptoris completes the main outline of the constellation, shining at magnitude 4.59 from a distance of around 140 light years.

Cetus, and to the east of the bright star Fomalhaut in Piscis Austrinus. Both Deneb Kaitos and Fomalhaut are included here as guides to locating this faint constellation, the whole of which is visible to observers south of latitude 50°N. However, because there are no bright stars in this group it may be difficult to make out for star-gazers at mid-northern latitudes unless the sky above their southern horizon is dark, clear and free of light pollution.

TOP: The strange spiral structure formed from material surrounding the red giant star R Sculptoris is thought to be caused by a hidden companion star.

OPPOSITE: One of Sculptor's showpieces is the spiral galaxy NGC 253.

Deneb Kaitos (in Cetus)

Fomalhaut (in Piscis Austrinus)

α

δ

γ

β

SCULPTOR

SCUTUM
The Shield

The small but well-defined shape of Scutum is bordered by Aquila to the northeast and Sagittarius to the south, the stars Lambda (λ) and 12 in Aquila being included here as a guide. Scutum is one of the smallest of the constellations, being ranked 84th in size, and is visible in its entirety from latitudes south of 74°N. The group was originally introduced by the Polish astronomer Johannes Hevelius in 1684 to honor King John III Sobiesci of Poland. The original name for the constellation was Scutum Sobiescianum (Sobiesci's Shield) although this has since been shortened to Scutum.

SCUTUM

THE STARS OF SCUTUM
The brightest star in Scutum is **Alpha (α) Scuti**, a magnitude 3.85 orange giant shining from a distance of around 200 light years.

Beta (β) Scuti lies at the northern end of Scutum, the light from this magnitude 4.22 yellow supergiant star having taken 900 years to reach us.

Magnitude 4.88 **Epsilon (ε) Scuti** is a yellow giant shining from a distance of 530 light years.

Delta (δ) Scuti shines at magnitude 4.70 from a distance of around 200 light years.

Located near the northeastern corner of Scutum is the magnitude 4.83 orange giant **Eta (η) Scuti**, the light from which set off towards us around 200 years ago.

Gamma (γ) Scuti lies at the southwestern corner of the group. Shining at magnitude 4.67 from a distance of 320 light years Gamma is one of the guide stars for locating the Eagle Nebula in the neighboring constellation Serpens Cauda (*see Serpens*) and the Omega Nebula in Sagittarius, which also borders Scutum (*see Sagittarius*).

THE SCUTUM STAR CLOUD
Scutum straddles the Milky Way and lies within a rich and star-packed area of sky, making it a wonderful target for the naked-eye backyard astronomer. This region is known as the Scutum Star Cloud

and is one of the richest parts of the Milky Way. There is no interstellar dust blocking out the light from the star fields beyond, and when viewed under dark, clear and moonless skies away from areas unspoiled by light pollution, this star-filled region of our Galaxy stands out well. The American astronomer Edward Emerson Barnard described it as being a *'Gem of the Milky Way'*, so take a look at the Scutum Star Cloud yourself and see if you agree with him!

THE WILD DUCK CLUSTER

Also known as the Wild Duck Cluster, the prominent open star cluster **Messier 11** (M11) or NGC 6705 lies at the northern edge of the Scutum Star Cloud. Discovered by the German astronomer Gottfried Kirch in 1681, M11 is located at a distance of around 6,000 light years and observation shows that the cluster contains almost 3,000 individual member stars. The Wild Duck Cluster derives its popular name from remarks made about it by William Henry Smyth, who likened its appearance to: ' *. . . a flight of wild ducks . . . a gathering of minute stars . . .* '.

Forming a small triangle with the nearby stars Beta and Eta, M11 shines at magnitude 6.3 and is fairly easy to locate. If you carefully sweep this area of sky with binoculars you will have little trouble tracking the cluster down, although you will need a small telescope in order to resolve some of its individual stars.

OPEN STAR CLUSTER M26

Situated a little way to the east of Delta, and lying in the same binocular field, is the open star cluster **Messier 26** (M26) or

NGC 6694. Credit for its discovery is usually given to Charles Messier, who recorded it in 1764, although it may have been seen by the French astronomer Le Gentil prior to then.

Shining with an overall magnitude of around 8 from a distance of 5,000 light years, M26 lies against the backdrop of the Milky Way and, unlike the nearby Wild Duck Cluster, may be difficult to detect. A line drawn from Alpha through Delta and on roughly half as far again will lead you

to M26 which will appear as a small concentration of stars against the background star fields. Provided the sky is dark, clear and moonless, binoculars will help you track it down and a small telescope should bring out some of the individual stars within the cluster.

ABOVE: The open star cluster M11 (NGC 6705) in Scutum is also known as the Wild Duck Cluster.

SERPENS
The Serpent

Serpens is unusual in that it is split into two sections, these being Serpens Caput (the Head) and Serpens Cauda (the Tail), although the two parts are regarded by astronomers as being a single constellation. Serpens represents a huge snake held by Ophiuchus (the Serpent Holder), the snake's head held in his left hand while his right hand is gripping the tail. The central part of Serpens is represented by the line of stars formed from Yed Prior, Yed Posterior, Upsilon (υ) Ophiuchi, Zeta (ζ) Ophiuchi and Sabik, all of which are actually in Ophiuchus. The constellation lies on or close to the celestial equator and can be viewed in its entirety from all inhabited regions of the world south of latitude 74°N, with at least portions of the group being visible worldwide.

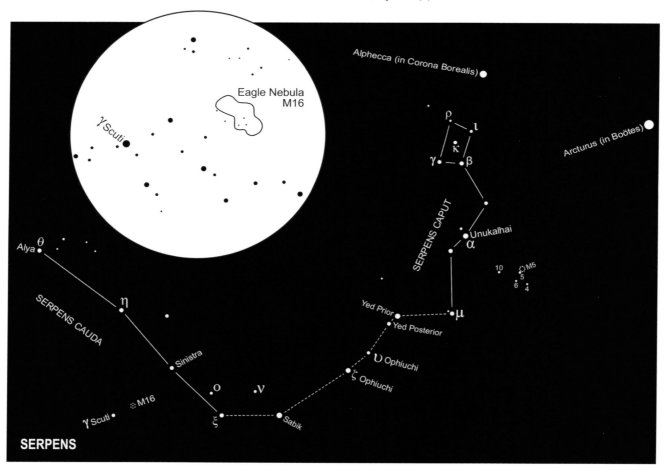

THE STARS OF SERPENS

The magnitude 2.63 orange giant **Unukalhai (α Serpentis)** is the brightest star in Serpens. Located a little to the south of the snake's head, the light from Unukalhai has taken 74 years to reach us. The name of this star is derived from the Arabic **'unuq al-hayya'** meaning 'the Serpent's Neck' reflecting its position in the constellation.

Serpens Cauda contains **Eta (η) Serpentis**, another orange star shining at magnitude 3.23 from a distance of 60 light years.

The light from magnitude 4.24 **Omicron (o) Serpentis** reaches us from a distance of 173 light years.

Shining from a distance of 105 light years is the white giant **Xi (ξ) Serpentis**, its magnitude 3.54 glow visible close to the southern border of Serpens.

Mu (μ) Serpentis is a blue giant shining at magnitude 3.54 from a distance of around 160 light years.

Both Serpens and Ophiuchus are best placed in or around July for observers in both hemispheres, the five stars depicting the head of the snake – Beta (β), Gamma (γ), Iota (ι), Kappa (κ) and Rho (ρ) Serpentis - forming a faint but distinctive quadrilateral to the south of the constellation Corona Borealis and a little to the east of the brilliant Arcturus in Boötes. Arcturus is included on the chart as a guide, along with Alphecca, the leading star of Corona Borealis. Once the five stars marking the head of the snake have been identified, the rest of the meandering form of Serpens can be picked out, given reasonably dark and clear skies and the use of a pair of binoculars.

DOUBLE STARS IN SERPENS

Alya (θ Serpentis) marks the tip of the snake's tail, its name taken from an Arabic word which actually refers to the tail of a sheep. Alya is a double star with magnitude 4.62 and 4.98 white components which are easily resolved in a small telescope.

Another double star is **Nu (ν) Serpentis** which lies close to the border with the neighboring constellation Ophiuchus and forms a small triangle with Omicron and Xi. A small telescope will resolve its magnitude 4.3 and 8.3 components.

GLOBULAR CLUSTER M5

Located on the northern edge of the small

field of faint stars 4, 5, 6 and 10 Serpentis, a little way to the southwest of Unukalhai, is the globular cluster **Messier 5** (M5) or NGC 5904. Discovered in 1702 by the German astronomer Gottfried Kirch, and shining with an overall magnitude of around 6 from a distance of 24,500 light years, M5 is one of the largest known globular clusters and contains well over 100,000 individual stars. M5 hovers at the threshold of naked-eye visibility and is easily picked up in binoculars, the cluster itself lying just to the northwest of the star 5 Serpentis. Small telescopes, or even binoculars with magnifications of 12x or more, may just help you to resolve some of the individual stars situated around the edges of this cluster.

LEFT: Serpens plays host to the unusually-shaped Red Square Nebula, which has been described as being one of the most symmetrical celestial objects ever discovered.

ABOVE: Hoag's Object in Serpens is an example of the class of objects known as ring galaxies.

201

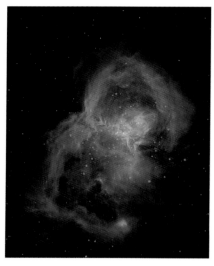

7,000 light years and was discovered by the Swiss astronomer Jean-Philippe Loys de Cheseaux in around 1746. It was independently discovered in 1764 by Charles Messier, who referred to the cluster being surrounded by a faint glow, an obvious reference to the nebulosity.

The Eagle Nebula shines with an overall magnitude of around 6.4, which puts it just below the limits of naked-eye visibility, the easiest way to track it down being to use the star Gamma (γ) Scuti, located at the southwestern corner of the

ABOVE LEFT: Globular cluster M5 (NGC 5904) in Serpens.

ABOVE: Located at a distance of around 1,600 light years is the star-forming region Westerhout 40 in Serpens Cauda.

OPPOSITE: The Eagle Nebula M16 (NGC 6611).

THE EAGLE NEBULA

Located a little way to the southeast of the star Nu we find **Messier 16** (M16) or NGC 6611, also known as the Eagle Nebula. This object takes the form of a large patch of diffuse nebulosity surrounding a bright open-star cluster. Both objects are visible through binoculars which will reveal some of the individual stars within the cluster as well as the nebulosity, which should appear as a faint hazy light surrounding the cluster. A small telescope will increase the number of visible cluster stars and larger instruments should bring out the nebulosity reasonable well.

Deriving its name from its appearance on photographic images, the Eagle Nebula lies at a distance of around

neighboring constellation Scutum, as a guide. Gamma Scuti is shown on the main chart, located to the southeast of the three stars Xi (ξ) and Eta (η) Serpentis (in Serpens Cauda) and Sinistra (in Ophiuchus). Start your search by locating Gamma Scuti from where you can use the finder chart (see page 200) to star hop your way to M16 through the field of fainter stars which lie in the immediate area.

ABOVE: The Pillars of Creation, the well-known star-forming region in the Eagle Nebula made famous by this incredible Hubble Space Telescope image.

SEXTANS
The Sextant

Lying to the south of Leo and to the east of Hydra is the tiny constellation Sextans. Taking the form of a small, dim and unimpressive triangle of faint stars, Sextans was introduced to star charts in the 17th century by the Polish astronomer Johannes Hevelius to commemorate the sextant, an instrument used by Hevelius to measure the positions of stars. The constellation can be seen to at least vaguely resemble the object it depicts!

Sextans lies on the celestial equator and parts of the group can be seen from anywhere, the whole of the constellation being visible from almost every inhabited

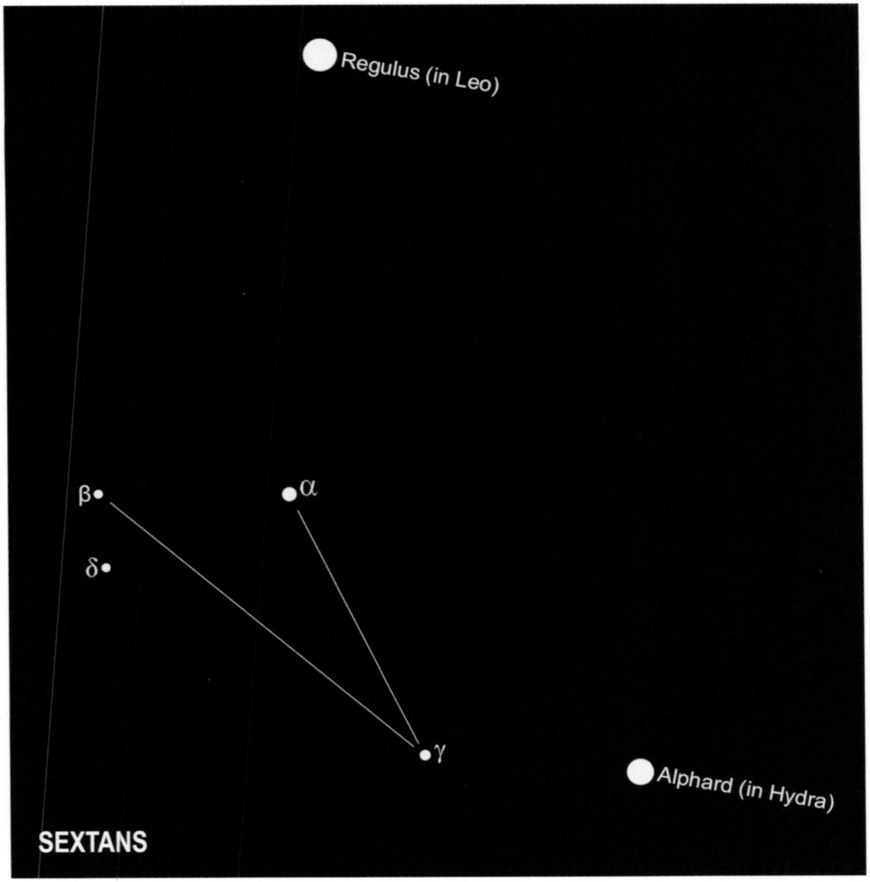

THE STARS OF SEXTANS

The brightest star in Sextans is the blue giant **Alpha (α) Sextantis** which shines with a magnitude of 4.48 from a distance of around 280 light years.

Slightly fainter is **Gamma (γ) Sextantis** which glows at magnitude 5.07, its light having taken 275 years to reach us.

Beta (β) Sextantis is almost identical in brightness to Gamma, the light from this magnitude 5.08 star reaching us from a distance of around 400 light years.

Rounding off the constellation is **Delta (δ) Sextantis**, a magnitude 5.19 star which lies around 320 light years away.

region of our planet. The constellation is shown here along with the nearby stars Regulus in Leo and Alphard in Hydra, both of which can be used as guides to locating it. None of the stars in Sextans have individual names and none is particularly bright. However, provided the sky is dark, clear and free of moonlight you might just be able to pick out this faint and obscure constellation with the naked eye, although a pair of binoculars would probably make your search easier!

Regulus (in Leo)

β•

δ•

•α

•γ

Alphard (in Hydra)

SEXTANS

TAURUS
The Bull

Located to the northwest of Orion is the prominent Taurus. The constellation represents Zeus who disguised himself as a white bull with the intention of abducting and carrying off Europa, the beautiful daughter of King Agenor of Phoenicia. Its leading star, Aldebaran, is easily located by following the line of stars in the Belt of Orion to the northwest, as shown here. In and around January, when suitably placed in the sky, Taurus can be seen in its entirety from every inhabited part of the world.

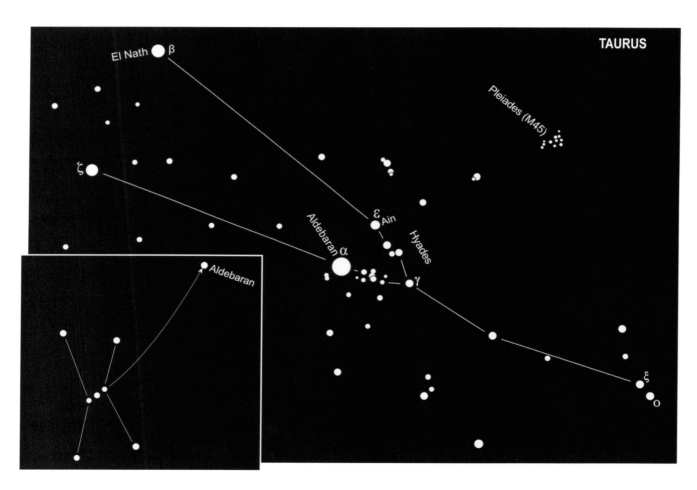

THE STARS OF TAURUS

Aldebaran (α Tauri) is an orange giant star shining at magnitude 0.87 from a distance of around 65 light years. With a diameter of more than 40 times that of the Sun, and a true luminosity of over 400 times that of our parent star, Aldebaran represents the southern eye of the celestial bull, the northern eye denoted by the yellow giant star Ain. Aldebaran takes its name from the Arabic *'al-dabaran'* meaning 'the Follower', alluding to the star 'following' the nearby Pleiades star cluster through the sky.

Deriving its name from the Arabic for 'the Butting One' is **El Nath (β Tauri)**. Shining with a magnitude of 1.65 from a distance of around 135 light years this is the second brightest star in Taurus and represents one of the horns of the celestial bull. El Nath appears to be associated with the neighboring constellation Auriga, although this is not the case *(see Auriga)*.

The other horn is depicted by **Zeta (ζ) Tauri**, the magnitude 2.97 glow of which reaches us from a distance of around 440 light years.

Magnitude 3.65 **Gamma (γ) Tauri** is a member of the Hyades star cluster *(see below)*, the light we are seeing from this yellow giant star having taken around 160 years to reach us.

Another yellow giant is **Omicron (o) Tauri**, located at the western end of Taurus and shining at magnitude 3.61 from a distance of 290 light years.

Immediately to the northeast of Omicron we find magnitude 3.73 **Xi (ξ) Tauri**, the light from which set off towards us 221 years ago.

The yellow giant star **Ain (ε Tauri)** derives its name from the Arabic *'ain al-thaur'* meaning 'the Bull's Eye' and denotes the northern eye of Taurus. Shining with a magnitude of 3.53 from a distance of 147 light years, Ain is one of the brighter members of the Hyades open-star cluster.

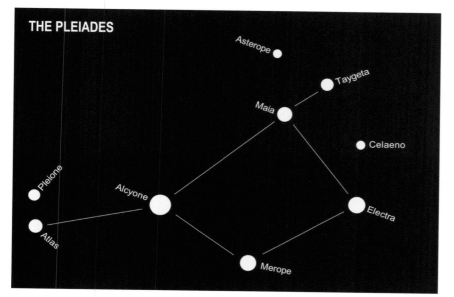

THE PLEIADES

Asterope

Taygeta

Maia

Celaeno

Pleione

Alcyone

Electra

Atlas

Merope

THE HYADES

The Hyades is a V-shaped collection of over 200 stars visible close to the star Aldebaran, as seen from Earth, and marking the head of Taurus. However, their apparent proximity to Aldebaran is somewhat misleading. The Hyades cluster lies at a distance of around 150 light years, over twice the distance of Aldebaran, which star only appears in the same line of sight as viewed from our planet. The Hyades is one of the nearest open star clusters and is thought to be over 600 million years old. Greek mythology identifies the Hyades as the daughters of Atlas and Aethra and half-sisters to the Pleiades. The Hyades had an elder brother Hyas, whom they loved dearly, but who was unfortunately killed while out hunting. So saddened were the sisters at

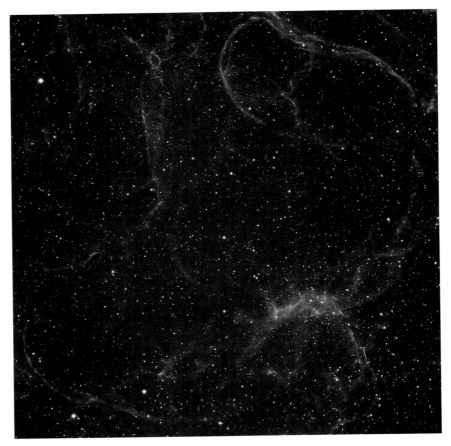

as seen from Earth. The naked eye will show around half-a-dozen cluster members while binoculars bring out many more. However, even a small telescope will reveal dozens of individual stars. Some of the stars in the Pleiades are enveloped in nebulosity. Visible only through large telescopes, this nebulosity is the remnant of the gas cloud from which the stars in the cluster were formed.

LEFT: The Hyades open star cluster.

BELOW: A color-enhanced image revealing the faint, wispy filaments of supernova remnant Sharpless 240 in Taurus.

OPPOSITE: The Pleiades (M45) open star cluster.

the loss of their brother that they died of grief, following which they were placed in the sky where we see them today.

THE PLEIADES

Somewhat more impressive than the Hyades are the Pleiades, catalogued as **Messier 45** (M45) by the French astronomer Charles Messier and without doubt the most famous open star cluster in the heavens. Commonly known as the Seven Sisters, the Pleiades were the daughters of Atlas and Pleione and half-sisters to the Hyades. This chart shows the named individual members of the Pleiades that are visible through binoculars.

Located at a distance of around 500 light years, the stars forming the Pleiades cluster are spread out over an area of space roughly equal to that of a full Moon

TELESCOPIUM
The Telescope

Located just to the north of Pavo is the tiny constellation Telescopium, devised by Nicolas Louis de Lacaille to commemorate the invention of the telescope following his stay in South Africa during 1751 and 1752. Telescopium is visible in its entirety from latitudes south of 33°N and takes the form of an extended circlet of faint stars, none of which are named and none of which are at all prominent.

THE STARS OF TELESCOPIUM

The magnitude 4.93 orange giant **Xi (ξ) Telescopii** marks the easternmost point of the main form of Telescopium. Xi lies just to the northwest of Peacock, the brightest star in Pavo and, once identified, the rest of Telescopium can be picked out. Peacock is included on the chart as a guide along with the nearby Alpha (α) Indi in the neighboring constellation Indus.

Alpha (α) Telescopii is the brightest star in the group, the light from this magnitude 3.50 star reaching us from a distance of over 250 light years.

Slightly fainter is **Zeta (ζ) Telescopii**, a magnitude 4.10 orange giant shining from a distance of around 125 light years.

Epsilon (ε) Telescopii is another orange giant star, the magnitude 4.52 glow of which reaches us from a distance of 415 light years.

Lambda (λ) Telescopii is a magnitude 4.85 white giant star located at a distance of just over 600 light years.

The magnitude 4.88 orange giant **Iota (ι) Telescopii** completes the main outline of Telescopium, its light having set off towards us around 370 years ago.

The two components of the optical double **Delta (δ) Telescopii**, located immediately to the east of Alpha, are resolvable in binoculars. The stars forming this double have similar magnitudes and both are blue-white, although the colors may not be easy to see. These two stars are situated at vastly differing distances, the magnitude 4.92 glow of Delta[1] (δ[1]) reaching us from just over 700 light years. This puts it a little over half as far away as magnitude 5.07 Delta[2] (δ[2]), the light from which set off towards us almost 1,200 years ago.

OPPOSITE: The globular cluster NGC 6584 in Telescopium shines from a distance of around 45,000 light years.

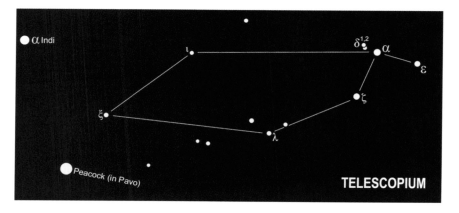

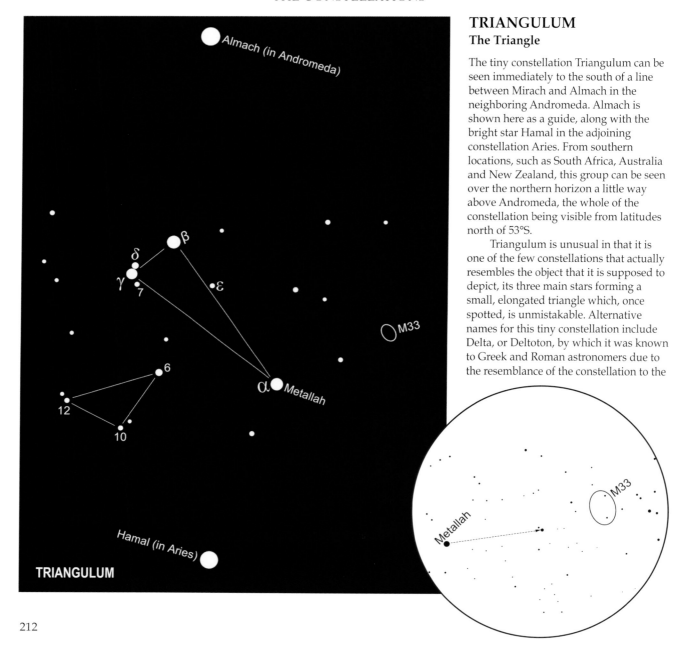

TRIANGULUM
The Triangle

The tiny constellation Triangulum can be seen immediately to the south of a line between Mirach and Almach in the neighboring Andromeda. Almach is shown here as a guide, along with the bright star Hamal in the adjoining constellation Aries. From southern locations, such as South Africa, Australia and New Zealand, this group can be seen over the northern horizon a little way above Andromeda, the whole of the constellation being visible from latitudes north of 53°S.

Triangulum is unusual in that it is one of the few constellations that actually resembles the object that it is supposed to depict, its three main stars forming a small, elongated triangle which, once spotted, is unmistakable. Alternative names for this tiny constellation include Delta, or Deltoton, by which it was known to Greek and Roman astronomers due to the resemblance of the constellation to the

THE STARS OF TRIANGULUM

The brightest star in Triangulum is the magnitude 3.00 white giant **Beta (β) Trianguli** which lies at a distance of around 125 light years.

Slightly fainter is **Metallah (α Trianguli)**, the name of this magnitude 3.42 star deriving from the Arabic **'al-muthal-lath'** meaning 'the Triangle'. The light from Metallah has taken a little over 60 years to reach us.

Magnitude 4.03 **Gamma (γ) Trianguli** shines from a distance of 112 light years. Just to the north of Gamma is **Delta (δ) Trianguli**, a yellow star shining at magnitude 4.84 from a distance of 35 light years. Immediately to the southwest of Gamma is **7 Trianguli**, the magnitude 5.25 glow of this star reaching us from a distance of 280 light years. The trio of stars Delta, Gamma and 7 Trianguli form an attractive little group when viewed through binoculars.

Located immediately to the north of the line between Beta and Metallah is **Epsilon (ε) Trianguli**, a magnitude 5.50 star shining from a distance of just under 400 light years.

Greek capital letter Delta (Δ). The shape of the group led to it being associated with Egypt and, in particular, with the delta of the Nile. The Latin author Hyginus recorded that the group was considered by some astronomers to have a shape not

RIGHT: The Triangulum Spiral Galaxy M33 (NGC 598).

unlike that of the island of Sicily, home of Ceres, the goddess of agriculture, the island originally being known as Trinacria due to its three promontories.

THE TRIANGULUM SPIRAL GALAXY

Located a short way to the northwest of the star Metallah, and shining from a distance of around three million light years, is the spiral galaxy **Messier 33** (M33) or NGC 598. Also known as the Triangulum Spiral Galaxy or Pinwheel Galaxy, M33 was discovered by the French astronomer Charles Messier in 1764, and described by him as: ' . . . *a whitish light of almost even brightness . . .'.* This object is the third largest member of the Local Group, a collection of galaxies of which our own Milky Way Galaxy and the Great Andromeda Galaxy are also members.

M33 has a tiny nucleus, or central region, together with huge sweeping spiral arms. This gives the galaxy a very low surface brightness and its pale glow is easily blotted out by moonlight or any form of light pollution. The overall magnitude of M33 is around 6 and, provided the sky is exceptionally dark and clear, you may just be able to make out the galaxy with the naked eye. Failing that, you should be able to track it down with binoculars or a small telescope by following the line of stars from Metallah as shown on the finder chart. The rule is to look for a faint and extensive patch of light rather than a more concentrated light source. Mounting your binoculars on a camera tripod will keep them steady and should help in picking out faint diffuse objects such as the Triangulum Spiral Galaxy.

TRIANGULUM MINUS

Located just to the southeast of Triangulum is a much smaller triangle of fainter stars. This is Triangulum Minus and was introduced to star charts by the Polish astronomer Johannes Hevelius in 1687, at which time he renamed the main constellation Triangulum Majus.

Often erroneously named Triangulum Minor, Triangulum Minus was created from the southern parts of his Triangula (the plural form of Triangulum) from the 5th-magnitude stars 6 Trianguli, 10 Trianguli and 12 Trianguli. This tiny triangle of faint stars is no longer recognized as an individual constellation on modern star charts.

LEFT: The Triangulum Spiral Galaxy imaged by the Hubble Space Telescope.

TRIANGULUM AUSTRALE
The Southern Triangle

The distinctively-shaped Triangulum Australe, known as the Southern Triangle to distinguish it from its northern counterpart Triangulum, takes the form of a small triangle of stars located a little way to the east of the prominent pair Alpha (α) and Beta (β) Centauri, both of which are included on the chart for guidance. Observers within or south of the Earth's equatorial regions can see the whole of Triangulum Australe, although the constellation is completely hidden from view to those north of latitude 30°N.

Triangulum Australe is one of the constellations introduced by Pieter Dirkszoon Keyser and Frederick de Houtman in the 1590s, its first appearance being on the celestial globe produced by

LEFT: Open star cluster NGC 6025 in Triangulum Australe.

Petrus Plancius in 1598. The stars in Triangulum Australe were later charted by the French astronomer Nicolas Louis de Lacaille, who gave the brightest stars in the constellation their Bayer designations *(see Glossary – Star Names).*

OPEN STAR CLUSTER NGC 6025

Triangulum Australe plays host to the open star cluster **NGC 6025**. This object

was discovered by Nicolas Louis de Lacaille, and is located on the northern edge of Triangulum Australe, near its border with the neighboring constellation Norma. NGC 6025 shines at magnitude 5.3 and contains around 30 individual stars. This is an easy target for binoculars and can be tracked down by following an imaginary line from Atria, through Delta, and extending it roughly half as far again.

THE STARS OF TRIANGULUM AUSTRALE

Atria (α Trianguli Australis) is the brightest star in Triangulum Australe. This magnitude 1.91 orange giant shines from a distance of 390 light years, the name of this star being a contraction of its Bayer reference Alpha Trianguli Australis.

The light from magnitude 2.83 **Beta (β) Trianguli Australis** has taken just 40 years to reach us.

Gamma (γ) Trianguli Australis completes the triangle, shining at magnitude 2.87 from a distance of around 180 light years.

Delta (δ) Trianguli Australis is a yellow supergiant, the magnitude 3.86 glow of which has taken around 600 years to reach us.

Orange giant **Epsilon (ε) Trianguli Australis** shines at magnitude 4.11 from a distance of 200 light years

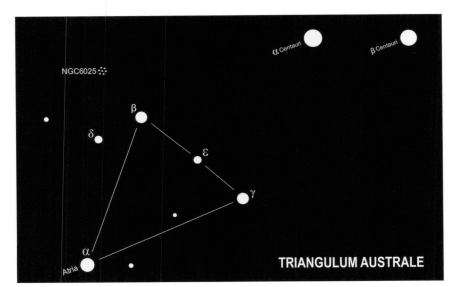

NGC6025

β

δ

ε

γ

α·Centauri

β·Centauri

α

Atria

TRIANGULUM AUSTRALE

TUCANA
The Toucan

The chart shows Tucana, together with the neighboring Hydrus *(see entry on 218)*, both of which lie in the general area of sky to the south of the bright star Achernar in Eridanus, which is shown here as a guide to locating these two groups. The main stars of Tucana form a small irregularly shaped circlet which can be seen in its entirety from anywhere in the southern hemisphere and from lower northern latitudes.

Tucana is another of the constellations introduced by Pieter Dirkszoon Keyser and Frederick de Houtman and depicts the well-known South American bird and which first appeared on the celestial globe produced by the Dutch celestial cartographer Petrus Plancius in 1598.

THE MAGNIFICENT 47 TUCANAE
Second only to Omega Centauri *(see Centaurus)* in terms of visual impact, the magnificent globular cluster **47 Tucanae** (NGC 104) is rated as being one of the finest globular clusters in the sky and is easily visible to the naked eye as a hazy, magnitude 4.5 star-like object. Located next to the Small Magellanic Cloud,

although lying at less than a tenth of the distance to that galaxy, 47 Tucanae shines from a distance of over 16,000 light years.

This splendid object was discovered by the French astronomer Nicolas Louis de Lacaille who described it as being: *'Like the nucleus of a fairly bright comet'.*

OPPOSITE: Globular cluster 47 Tucanae (NGC 104).

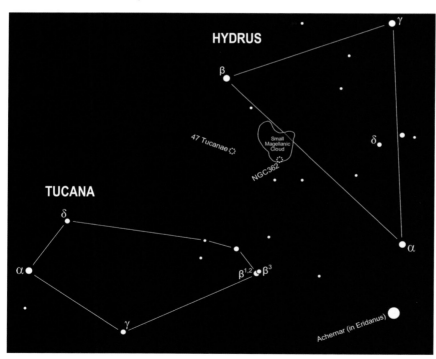

47 Tucanae is a huge system, with a diameter of 120 light years and a true total luminosity of over a quarter of a million times that of the Sun. Binoculars with magnifications of 10x or more, or a small telescope, will resolve individual stars at the cluster's edge, the degree of resolution increasing rapidly as larger instruments are turned towards the cluster.

GLOBULAR CLUSTER NGC 362
Discovered in August 1826 by the Scottish astronomer James Dunlop, the globular cluster **NGC 362** shines with an overall magnitude of 6.4, putting it just below naked eye visibility. Lying against the backdrop of star fields at the edge of the Small Magellanic Cloud, binoculars will be needed in order to track it down, and small telescopes will start to resolve individual stars within the cluster. Shining from a distance of around 28,000 light years, NGC 362 is less condensed and somewhat fainter than 47 Tucanae.

HYDRUS
The Little Water Snake

Bordering Tucana is the constellation Hydrus, this being another of the groups introduced by Pieter Dirkszoon Keyser and Frederick de Houtman in the 1590s. Taking the form of a small triangle made up from its three principal stars, Hydrus contains no particularly bright stars, and none have individual names. As is the case with Tucana, Hydrus can be seen in its entirety from regions around the equator and locations further to the south.

THE SMALL MAGELLANIC CLOUD

There are a number of objects in Tucana of interest to the backyard astronomer, not least of which is the Small Magellanic Cloud (SMC), a dwarf irregular galaxy and a member of the Local Group of Galaxies *(see Glossary)*. Straddling the border between Tucana and the adjoining constellation Hydrus, the SMC is located at a distance of 190,000 light years and has a diameter of around 7,000 light years. Early navigators referred to the Small Magellanic Cloud (and the nearby Large Magellanic Cloud - *see Dorado*) as the Cape Clouds, in that they were a pair of prominent celestial objects seen as they approached the Cape of Good Hope. They were also known to many ancient civilizations, although it was only following accounts of the circumnavigation of the Earth by the Portuguese explorer Ferdinand Magellan, in the early 16th century, that their presence became more widely known. Magellan's contribution eventually

THE STARS OF HYDRUS
Beta (β) Hydri is one of our stellar neighbors, the light from this magnitude 2.82 star reaching us from a distance of just 24.3 light years.

Slightly fainter and more remote is **Alpha (α) Hydri**, the light from this magnitude 2.86 star having taken 72 years to reach us.

The main form of Hydrus is completed by the magnitude 3.26 red giant **Gamma (γ) Hydri**, the light from which set off towards us a little over 200 years ago.

Delta (δ) Hydri shines at magnitude 4.08 from a distance of around 140 light years.

resulted in both objects being named in his honor.

The Small Magellanic Cloud has a fairly low surface brightness and is best viewed from dark locations away from any light pollution. Visible to the naked eye as a hazy, elongated patch of light, binoculars show it quite well, and careful sweeping of the Small Magellanic Cloud with a telescope or good binoculars will reveal the presence of a number of nebulae and star clusters.

ABOVE LEFT: Globular cluster NGC 362 in Tucana.

OPPOSITE: Several star-forming regions are evident in this Hubble Space Telescope image of the Small Magellanic Cloud.

URSA MAJOR
The Great Bear

One of the most famous and easily-recognizable star patterns is the Plough which, shaped rather like a gigantic heavenly spoon, is probably the best-known asterism in the sky, being part of the much larger constellation Ursa Major. Although the other stars in the Ursa Major are all comparatively faint, the Plough itself, which marks the hindquarters and tail of the bear, stands out quite well. Ursa Major is visible in its entirety from locations north of latitude 17°S, making it a predominantly northern constellation. When viewed from mid-northern latitudes the Plough, together with the rest of Ursa Major, is located fairly well up in the northeastern sky during evenings in March and April. As far as observers in the southern hemisphere are concerned, this is the best time to see Ursa Major, although even then it is not easy. From most of South Africa, Australia and southern South America the parts of Ursa Major that rise at all are very low in the northern sky and from locations further south very little if any of the constellation can ever be seen.

Ursa Major is one of those constellations that actually resembles the object or character that it depicts and checking out the group as a whole you will see that the legs and head of the Great Bear are clearly represented. When seen from mid-northern latitudes the Plough, together with the rest of Ursa Major, is located at or near the overhead point during spring, and as the year wears on drops down into the high northwest,

URSA MAJOR

Dubhe (α Ursae Majoris) derives its name from the Arabic *'al-dubb'* meaning 'the Bear' and is a yellow giant star shining at magnitude 1.81 from a distance of 123 light years.

Taking its name from the Arabic for 'The Groin (of the Great Bear)' magnitude 2.34 **Merak (β Ursae Majoris)** lies at a distance of 80 light years.

Shining from a distance of 80 light years is magnitude 3.32 **Megrez (δ Ursae Majoris)**, a star which derives its name from an abbreviation of the Arabic for 'The Root of the Tail'.

The light from magnitude 2.41 **Phekda (γ Ursae Majoris)** set off on its journey towards us 83 years ago. Phekda derives its name from the Arabic *'fakhidh al-dubb al-akbar'* meaning 'The Thigh (of the Great Bear)'.

Benetnash (η Ursae Majoris) marks the tip of the Great Bear's tail, its magnitude 1.85 glow reaching us from a distance of 104 light years

Magnitude 1.76 **Alioth (ε Ursae Majoris)** lies at a distance of 83 light years

The Great Bear's hind feet are depicted by the pair of stars **Tania Borealis (λ Ursae Majoris)** and **Tania Australis (μ Ursae Majoris)**. Tania Borealis shines at magnitude 3.45 from a distance of 137 light years, the light from magnitude 3.06 red giant Tania Australis having taken 230 years to reach us.

The Great Bear's front paws are marked by the magnitude 3.57 white giant star **Kappa (κ) Ursae Majoris** and magnitude 3.12 **Talitha (ι Ursae Majoris)**. Talitha shines from a distance of 47 light years putting it considerably closer than Kappa, the light from which reaches us from a distance of around 355 light years.

Muscida (o Ursae Majoris) is a magnitude 3.35 yellow giant located at a distance of around 180 light years.

eventually ending up over the northern horizon in autumn. This annual progression of the Great Bear around the northern sky gave rise to one of the many legends attached to the Plough. The Native Americans identified the four stars in the 'bowl' of the Plough as a bear, which was being chased around the sky by three hunters, represented by the three stars in the Plough 'handle'. The bear is pursued until autumn when it is

eventually caught. From mid-northern latitudes the group can be seen low over the northern horizon, the blood from the bear supposedly dripping down onto the earth below, turning the leaves brown.

OPPOSITE: Discovered by the American astronomer Edwin Coddington in 1898, and shining from a distance of around 12 million light years, Coddington's Nebula (IC 2574) is a dwarf irregular galaxy in Ursa Major.

ALCOR, MIZAR AND SIDUS LUDOVICIANA

A close look at the star in the middle of the Plough 'handle' will show that this is actually a pair of stars comprising **Mizar (ζ Ursae Majoris)** and **Alcor (80 Ursae Majoris)**. Magnitude 2.23 Mizar is the brighter of the two, outshining its magnitude 3.99 companion Alcor. If you have really keen eyesight, and the sky is very dark and clear, you may be able to resolve this pair with the naked eye, although binoculars may be needed to bring the pair out well. Powerful binoculars, or a small telescope, will reveal a much fainter 8th magnitude star forming a triangle with Alcor and

Mizar. This star is known as Sidus Ludoviciana, being named as such by the eccentric German astronomer Johann Georg Liebknecht in December 1722. Liebknecht spotted the star in his telescope and, thinking he had discovered a new planet, named it after his sovereign and patron the Landgrave Ludwig of Hessen-Darmstadt. He went on to publish details of his 'find' although the discovery proved to be erroneous. Sidus Ludoviciana turned out to be nothing more than an ordinary star, although the unwieldy name he came up with for it remained, and is a permanent reminder of

Liebknecht's fruitless and misguided efforts. Sidus Ludoviciana is not actually associated with Alcor and Mizar and only happens to lie in the same line of sight.

Both Alcor and Mizar lie at a distance of around 82 light years, considerably closer than Sidus Ludoviciana whose light has taken around 400 years to reach us. If you check out Mizar with a small telescope you will notice a faint 4th magnitude companion lying close by, his star being first seen in around 1650 by the Italian astronomer Giovanni Battista Riccioli.

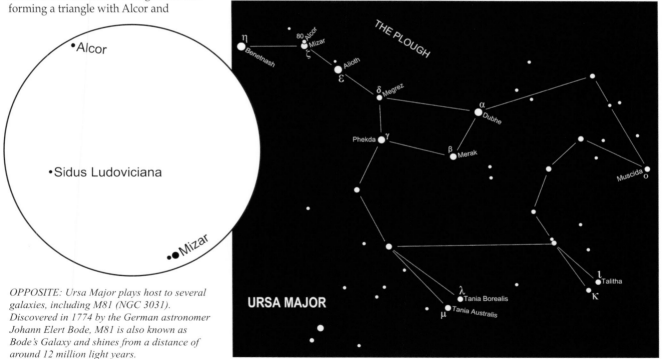

OPPOSITE: Ursa Major plays host to several galaxies, including M81 (NGC 3031). Discovered in 1774 by the German astronomer Johann Elert Bode, M81 is also known as Bode's Galaxy and shines from a distance of around 12 million light years.

URSA MINOR
The Little Bear

Ursa Minor is one of the oldest star patterns, appearing in the catalogue of constellations compiled during the second century by the Greek astronomer Ptolemy, and is a familiar sight to observers at mid-northern latitudes. When suitably placed in the sky, Ursa Minor can be seen in its entirety from anywhere north of the equator, although from latitudes below around 25°S the whole of the constellation is permanently hidden from view.

Ursa Minor is by no means the most prominent constellation in the sky, although it stands out reasonably well due to the fact that the area around it is devoid of bright stars. It is easily located by using the two end stars in the 'bowl' of the Plough as pointers, a line from Merak through Dubhe eventually leading you to the Polaris, also known as the Pole Star. This is the brightest star in Ursa Minor, the constellation as a whole bearing a resemblance to the Plough and stretching away southwards from its leading star Polaris.

Polaris marks the position of the north celestial pole (see Glossary), its southern counterpart being located in the tiny constellation Octans (the Octant). However, unlike Polaris and the north celestial pole, there is no particularly bright star marking the position of the south celestial pole (see Octans).

Polaris is lined up with the Earth's axis of rotation, the consequence of which is that it appears to remain stationary as the Earth turns on its axis, with all the other stars appearing to go round it every 24 hours. This situation forms the basis of

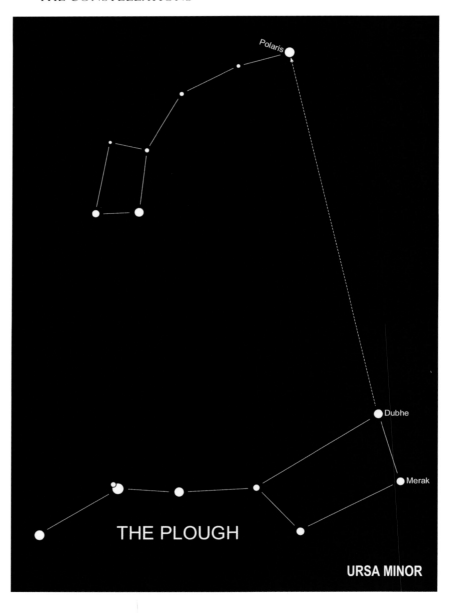

URSA MINOR

Polaris (α Ursae Minoris) shines at magnitude 2.02 from a distance of 430 light years. Although this star does not appear to be particularly bright, its true luminosity is around 2,500 times that of our Sun, and only appears faint due to the fact that it lies at such a great distance from us.

Kochab (β Ursae Minoris) and **Pherkad (γ Ursae Minoris)** are known collectively as the Guardians of the Pole due to the fact that they appear to circle around Polaris as the Earth rotates. Kochab is an orange giant, its magnitude 2.07 glow reaching us from a distance of just 130 light years, making is one of the closest stars in Ursa Minor. When viewed through binoculars the orange tint of Kochab contrasts well with the white of Polaris. The magnitude 3.00 white giant Pherkad lies at a distance of around 490 light years. This star is sometimes known as Pherkad Major to distinguish it from nearby 11 Ursae Minoris (Pherkad Minor) *(see below)*.

Pherkad Minor (11 Ursae Minoris) is an orange giant shining at magnitude 5.02 from a distance of 398 light years. Pherkad and Pherkad Minor can be separated with the naked eye under really clear, dark skies and were alluded to by the astronomer Joseph Henry Elgie when he said: *'The very small star close to Gamma is visible to me but seldom. I can just see it at this moment – a shivering moment, for the piercingly cold wind hums about me the while. The Form of the Lesser Bear (Ursa Minor) always reminds me of the Great Bear turned inside out.'*

Yildun (δ Ursae Minoris) shines at magnitude 4.35 from a distance of 172 light years.

The magnitude 4.29 glow of **Zeta (ζ) Ursae Minoris** reaches us from a distance of around 370 light years.

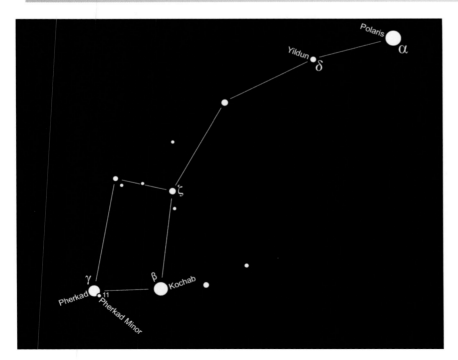

an Arabic legend which identifies Polaris as a notorious villain who was cast into the sky upon his death. His coffin is represented by the Pole Star and was placed at the northernmost part of the sky, all the other stars keeping their distance as they travel around the heavens.

Even a cursory glance at the celestial bears *(see also Ursa Major)* will reveal that they both have long tails, unlike their earthly counterparts. According to legend, Ursa Major represents the legendary maiden Callisto who was so beautiful that the Roman goddess Juno, the wife of Jupiter, became jealous and turned her into a bear. Years later, Callisto's son Arcas almost killed the bear while out hunting. Jupiter rescued the situation by turning Arcas into a bear and, in order to render both Callisto and Arcas safe from Juno's clutches, grabbed both animals by their tails and swung them high up into the sky where we see them today, the tails of both animals having become stretched in the process!

225

VELA
The Sail

Located immediately to the north of Carina and representing the sails of the dismantled ship Argo Navis, the whole of Vela can be observed from anywhere south of latitude 33°N, although the entire constellation is hidden from view to those located at latitudes north of Winnipeg, Canada and London, England.

THE OMICRON VELORUM CLUSTER

The brightest of the star clusters found in Vela is **IC 2391** which is readily visible to the naked eye just to the north of the star Delta and appears to have been first recorded as long ago as the tenth century by the Persian astronomer Al-Sufi. Independently discovered by the French astronomer Nicolas Louis de Lacaille in 1751, IC 2391 is also known as the Omicron Velorum Cluster, its brightest member being the magnitude 3.60 blue

THE STARS OF VELA

In reality a four-star system, **Gamma (γ) Velorum** is the brightest star in Vela. Two of the components, shining at magnitudes 1.9 and 4.2, are visible as a double star and can be resolved either in high-magnification binoculars or a small telescope.

The third brightest star in Vela is the magnitude 2.23 supergiant **Suhail (λ Velorum)**. Located at a distance of well over 500 light years, the distinctly orange hue of this star can be seen quite well in binoculars.

Located near the eastern border of Vela is the yellow giant star **Mu (μ) Velorum**, the magnitude 2.69 glow of which reaches us from a distance of around 115 light years.

Psi (ψ) Velorum marks the northern border of Vela, shining at magnitude 3.60 from a distance of 61 light years.

Magnitude 2.47 **Markeb (κ Velorum)** is situated near the southern border of Vela. Probably deriving its name from the Arabic **'markab'** meaning 'a ship or any vehicle', the light from this star set off towards us around 570 years ago.

Delta (δ) Velorum shines at magnitude 1.93 from a distance of 80 light years.

Markeb and Delta, together with Iota (ι) and Epsilon (ε) from the neighboring constellation Carina, form the famous asterism known as the False Cross *(see Carina)*.

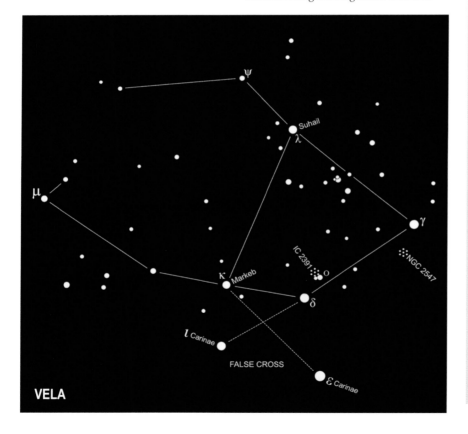

VELA

giant **Omicron (o) Velorum**. Shining with an overall magnitude of 2.5 from a distance of 500 light years, IC 2391 contains around 30 member stars and is an excellent target for binoculars.

OPEN STAR CLUSTER NGC 2547
Also discovered by Lacaille, the open cluster **NGC 2547** is somewhat fainter than IC 2391 shining with an overall magnitude of 4.7 from a distance of

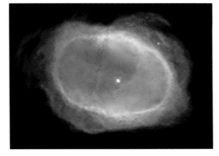

around 1,500 light years. Located a little way to the south of the star Gamma, NGC 2547 can be seen with the naked eye and is worth seeking out with binoculars.

LEFT: Planetary nebula NGC 3132, also known as the Eight-Burst or Southern Ring Nebula.

BELOW: Open star cluster NGC 2547 in Vela.

VIRGO
The Virgin

The long and sprawling pattern of stars that forms the constellation Virgo manifest themselves as a conspicuous Y-shape straddling the celestial equator, a position which ensures that the whole of the constellation is visible from every inhabited part of the world south of latitude 68°N. The brightest star in Virgo is Spica, and observers in the northern hemisphere can track this star down by following the curve of stars in the handle of the Plough (see Ursa Major) southwards. The first bright star you will come across

is Arcturus in the constellation Boötes, although extending the line as far again will lead you to Spica. Those in the southern hemisphere, who may not be able to see the Plough, can identify Spica as one of three bright stars forming a prominent triangle in the northern sky, the other two being Arcturus in Boötes and Denebola in Leo.

The ancient Greeks associated Virgo with Ceres, the goddess of the harvest, Spica depicting an ear of wheat held in Virgo's left hand. Egyptian mythology suggests that the misty band of light we know as the Milky Way was created by Virgo throwing millions of wheat heads

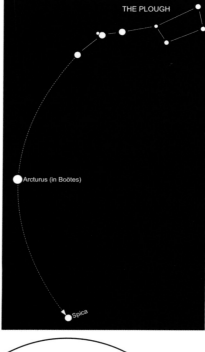

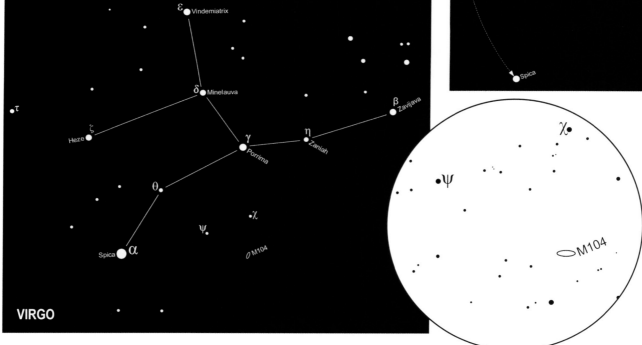

VIRGO

Spica (α Virginis) is the brightest member of Virgo, the light from this magnitude 0.98 blue-white star having taken around 250 years to reach us. Spica takes its name from its Roman title meaning 'the Ear of Grain' and is one of the four bright stars forming the prominent asterism known as the Diamond of Virgo *(see Coma Berenices)*, the other three being Arcturus in Boötes, Denebola in Leo and Cor Caroli in Canes Venatici.

Porrima (γ Virginis) marks the central point of Virgo, this magnitude 2.74 star shining from a distance of 38 light years.

Zavijava (β Virginis) is a yellow-white star located at the western end of Virgo and shining at magnitude 3.59 from a distance of 36 light years.

Located at a distance of around 200 light years, the orange-red tint of the magnitude 3.39 red giant **Minelauva (δ Virginis)** can be seen in binoculars.

Located near the northern border of Virgo, the magnitude 2.85 **Vindemiatrix (ε Virginis)** is a yellow giant situated at a distance of 110 light years. Vindemiatrix takes its name from the Latin for 'the Grape Gatherer', alluding to the fact that the star was seen to rise just before the Sun at around the time of the annual harvest.

The light from magnitude 3.38 white star **Heze (ζ Virginis)** reaches us from a distance of around 74 light years.

The remaining two stars that make up the main outline of Virgo are magnitude 4.38 **Theta (θ) Virginis**, the light from which has taken around 310 years to reach us, and **Zaniah (η Virginis)**, a magnitude 3.89 star located at a distance of 265 light years.

Tau (τ) Virginis shines at magnitude 4.23 from a distance of 220 light years. Tau is located to the east-northeast of Heze and a close look at this star with either a small telescope or powerful binoculars will reveal a 9th magnitude companion star lying nearby.

---

up into the heavens. Perhaps worryingly, the Greek philosopher Pliny believed that if ever a comet was seen in the constellation, great misfortune would befall all females here on Earth.

THE SOMBRERO HAT GALAXY

Located to the south of Porrima, and lying almost on the border between Virgo and the neighboring constellation Corvus (the Crow), is the galaxy **Messier 104** (M104) or NGC 4594. Located at a distance of around 30 million light years, this object was discovered in 1781 by the French astronomer Pierre Méchain who, when viewing it through his small

RIGHT: Markarian's Chain is a sweep of galaxies which lie along a curved line as seen from Earth and which form part of the Virgo Cluster of galaxies.

telescope, described it as: ' . . . *a nebula above Corvus which did not appear to contain a single star'.*

When William Herschel turned his considerably larger telescope towards M104 he noted the presence of ' . . . *a dark interval of stratum separating the nucleus and the general mass of the nebula . . .'.* It is this feature that has led to M104 being known as the 'Sombrero Hat Galaxy', the name deriving from the appearance of the galaxy when viewed through large telescopes. From our position in space we see M104 edge-on and large telescopes reveal a conspicuous lane of dark, dusty material spread out along its main plane, with the bright central regions of the galaxy projecting outwards to either side.

Although a large telescope is required to see this dust lane, the galaxy itself may be glimpsed with only moderate optical aid. M104 has an overall magnitude of around 8.5 and can be seen to form a small triangle with the two faint stars Psi (ψ) and Chi (χ) Virginis, which are located in the area of sky just to the south of Porrima. Once these two guide stars are identified you can 'star hop' your way to the Sombrero Hat Galaxy by carefully following the patterns of faint stars shown on the finder chart *(see page 228)*. Providing the sky is dark, clear and moonless, and you have plenty of patience, you should manage to locate M104 with either a good pair of binoculars or a small telescope, bearing in mind that you are looking for a patch or sliver of pale light rather than a star-like point.

RIGHT: The Sombrero Hat Galaxy M104 (NGC 4594) in Virgo.

VOLANS
The Flying Fish

Situated immediately to the south of Carina, and shown here along with the two neighboring bright stars Miaplacidus and Epsilon (ε) Carinae as a guide to location, the tiny constellation Volans lies quite close to the south celestial pole and is one of the constellations introduced by Pieter Dirkszoon Keyser and Frederick de Houtman in the 1590s.

This group was originally known as Piscis Volans, and appeared on the celestial globe produced by Petrus Plancius in 1598, following which it was depicted in the star atlas *Uranometria*, produced in 1603 by the German astronomer and celestial cartographer Johann Bayer. The whole of Volans is visible from locations to the south of latitude 15°N. There are no legends attached to the constellation and none of its stars are named.

OPPOSITE: Located at a distance of around 50 million light years is the spiral galaxy NGC 2442 / 2443, also known as the Meathook Galaxy, in Volans.

THE STARS OF VOLANS

The light from magnitude 4.00 **Alpha (α) Volantis** has taken around 125 years to reach us. Alpha lies on the border of Volans, a little to the north of Miaplacidus in the neighboring constellation Carina, and from here the rest of the constellation can be picked out.

The brightest star in Volans is the magnitude 3.77 orange giant **Beta (β) Volantis** which shines from a distance of a little over 100 light years.

Gamma (γ) Volantis is a pretty double star, the magnitude 3.78 and 5.68 yellowish components of which can be resolved in small telescopes.

Zeta (ζ) Volantis is a magnitude 3.93 yellow giant star located at a distance of 140 light years.

The main outline of Volans is completed by the magnitude 3.97 white supergiant **Delta (δ) Volantis**, the light from which has taken around 740 years to reach our planet, and the slightly fainter **Epsilon (ε) Volantis**, the magnitude 4.35 glow of which reaches us from a distance of around 550 light years.

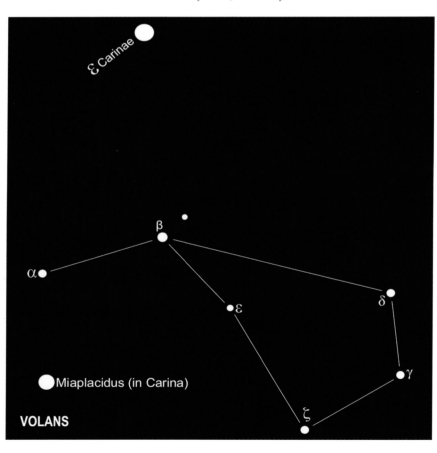

ε Carinae

β

α

ε

δ

Miaplacidus (in Carina)

ζ

γ

VOLANS

VULPECULA
The Fox

Taking the form of a zig-zag line of faint stars, and located immediately to the south of the constellation Cygnus, Vulpecula is one of the constellations introduced by the Polish astronomer Johannes Hevelius. As guides to locating this faint group, the chart includes the nearby bright star Albireo (in Cygnus), together with the main outline of the tiny constellation Sagitta (the Arrow), which borders Vulpecula to the south.

Vulpecula lies a little way to the north of the celestial equator, and the whole of

the constellation is visible from all locations north of latitude 60°S. However, because the brightest stars in this group are by no means conspicuous, you may need a little patience picking Vulpecula out against the background star fields.

Hevelius originally named this constellation Vulpecula et Anser ('the little fox and the goose') and depicted it as a fox holding a goose in its jaws. The Fox and the Goose were then split into separate groups, but were later reunited into a single constellation, the Goose, now represented by the red giant star **Anser (α Vulpeculae)**, the brightest star in Vulpecula. Shining at magnitude 4.44

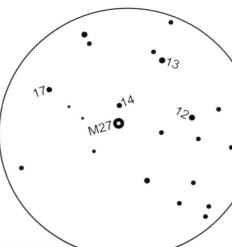

from a distance of 297 light years, the color of this star can be distinguished in binoculars, which will also reveal a magnitude 5.82 companion lying close by. This pairing makes an easy double-star target for binoculars, although the lining-up of the two stars is purely due to the fact that they both happen to lie in roughly the same line of sight as seen from our planet, the fainter of the two being situated nearly twice as far away as Anser.

THE DUMBBELL NEBULA

Located at a distance of around 1,350 light years, and shining with an overall magnitude of 7.4, **Messier 27** (M27) or NGC 6853 was the first planetary nebula to be discovered, and is considered to be the finest object of its type in the sky. Also

THE STARS OF VULPECULA

1 Vulpeculae lies to the southwest of Anser, the light from this magnitude 4.76 star having taken just over 800 years to reach us.

The magnitude 4.57 blue giant **13 Vulpeculae** is somewhat closer, shining from a distance of 333 light years.

Completing Vulpecula's zig-zag pattern are the magnitude 4.66 white giant **15 Vulpeculae** which, at a distance of 236 light years, is a little nearer to us than magnitude 4.50 orange giant **23 Vulpeculae**, the light from which has taken around 340 years to reach us.

known as the Dumbbell Nebula, it was first seen by Charles Messier in 1764 and derives its name from its appearance when seen through a telescope. The English astronomer Thomas William

Webb described it as being ' . . . *two oval hazy masses in contact . . .* ', his

ABOVE: The Dumbbell Nebula M27 (NGC 6853) in Vulpecula.

235

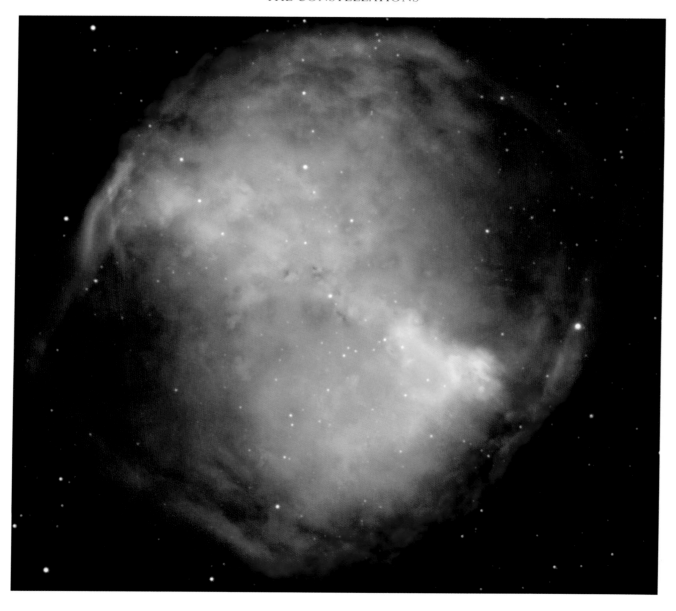

contemporary and countryman Edwin Dunkin going even further, portraying M27 as being a: '. . . *nebula shaped like a dumb-bell or hour-glass . . . in the constellation Vulpecula . . . consist(ing) of two luminous symmetrical patches, joined together by a narrow isthmus . . .* '. Both these descriptions illustrate the Dumbbell Nebula very well.

M27 is an easy target for binoculars and small telescopes and is well worth seeking out. If you locate the two stars 13 and 14 Vulpeculae, you should then glimpse the Dumbbell Nebula immediately to the south of 14, as shown on the finder chart. Providing the sky is dark and clear, binoculars should reveal the Dumbbell Nebula as a small, fuzzy patch of light and closer examination with even a small telescope should start to reveal the distinctive shape for which this object is famous.

THE COATHANGER CLUSTER

Located near the southwestern corner of Vulpecula is the Coathanger Cluster, perhaps the most distinctive object in the constellation. Located a little to the north of the two end stars, in the neighboring constellation Sagitta, this object was first seen and catalogued by the Persian astronomer Al-Sufi in the tenth century, who described it as '. . . *a little cloud situated to the north of the two stars of the notch of Sagitta*'. The crystal clear skies that Al-Sufi worked under enabled him to see the cluster with the naked eye and, if your sky is really dark, clear and moonless, you might also be able to make out the Coathanger with the unaided eye as a tiny fuzzy patch of light.

A pair of binoculars will enable you to track the Coathanger down. To find it, first of all locate the tiny constellation Sagitta, the main outline of which is shown here, and follow a line northwards from Beta (β), through Sham and on to the star 9 in Vulpecula. The distinctive shape of the Coathanger will then be visible just a short way to the west of 9, the main members of the cluster being depicted on the chart.

The Coathanger is a must-see for the backyard astronomer. Arranged around the stars 4, 5 and 7 in Vulpecula, the cluster's distinctive shape is brought out well in the wide field view of binoculars, a row of six stars forming the 'bar' of the Coathanger and four more to the south of

these forming the 'hook'. It should be pointed out that the Coathanger isn't a true star cluster, but rather a chance alignment of stars whose distances from us range from a little over 200 light years to nearly 1,000 light years. However, this doesn't make the Coathanger any the less interesting to look at, and observers who spot it for the first time seldom fail to be impressed by its unusual appearance.

OPPOSITE: A color-enhanced image of the Dumbbell Nebula.

BELOW: Although not a true cluster and formed from a chance-alignment of unrelated stars, the distinctive Coathanger is well worth seeking out.

GLOSSARY

ALTITUDE
The angular distance of a star or other object above the horizon. For example, if a star is located at the zenith, or overhead point, its altitude is 90° and if it is on the horizon, its altitude is 0°.

ASTERISM
A grouping or collection of stars, located within a constellation, that form an apparent and distinctive pattern.

Examples include the False Cross (formed from stars in Carina and Vela); the Square of Pegasus (in Pegasus) and the Summer Triangle (formed from stars in Lyra, Cygnus and Aquila).

AVERTED VISION
Averted vision is a useful technique for observing faint objects which involves looking slightly to one side of the object under observation. By doing so you allow the light emitted by the object to fall on a more sensitive part of the retina. Although you're not looking directly at the object, it is surprising how much more detail comes into view. This technique is also useful when observing double stars which have components of greatly contrasting brightness. Although direct vision may not reveal the glow of a faint companion star in the glare of a much brighter primary, averted vision may well bring the fainter star into view.

BINARY STAR
See Double Star

CELESTIAL EQUATOR
A projection of the Earth's equator onto the celestial sphere, equidistant from the celestial poles and dividing the celestial sphere into two hemispheres.

CELESTIAL POLES
The north (and south) celestial poles are points on the celestial sphere directly above the north and south terrestrial poles around which the celestial sphere appears to rotate and through which extensions of the Earth's axis of rotation would pass *(see Octans and Ursa Minor)*.

The north celestial pole would be seen directly overhead when viewed from the North Pole, and the south celestial pole directly overhead when seen from the South Pole. Consequently, the north celestial pole lies in the direction of north when viewed from elsewhere on the Earth's surface and the south celestial pole lies in the direction of south when viewed from other locations.

The Milky Way is seen here passing through the region of sky containing the asterism known as the Summer Triangle, which is formed from the three bright stars Vega (in Lyra), Deneb (in Cygnus) and Altair (in Aquila) and which occupies the left half of the image. By comparing this picture with the constellation chart shown on the opposite page, which depicts the area of sky around the Summer Triangle, you should be able to identify the three stars forming the Summer Triangle. You should also be able to pick out one or two neighboring constellations, including Delphinus and Sagitta (two tiny groups located near the star Altair in Aquila, just below center of image) and the head of Draco (visible to the upper left of Vega).

WHAT STAR?

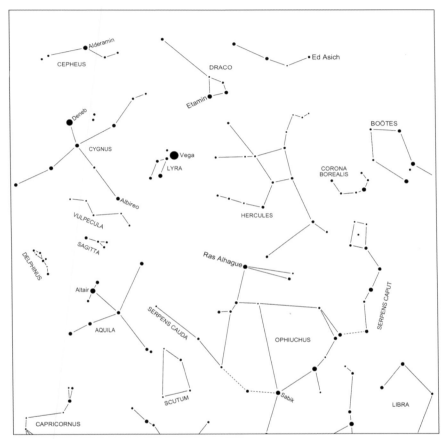

Section of a star chart depicting the stars and constellations in and around the Summer Triangle (see also image on opposite page).

The south celestial pole is located in the tiny constellation Octans (the Octant) although there is no particularly bright star marking its position, unlike the north celestial pole which lies in the constellation Ursa Minor (the Little Bear), the position of which is marked by the relatively bright star Polaris.

CELESTIAL SPHERE

The imaginary sphere surrounding the Earth on which the stars appear to lie.

CIRCUMPOLAR STAR

A star which never sets from a given latitude. When viewing the sky from either the North or South Pole, all stars will be circumpolar, although no stars are circumpolar when viewed from the equator.

CONSTELLATION

A constellation is an arbitrary grouping of stars forming a pattern or imaginary picture on the celestial sphere. Many of these have traditional names and are associated with folklore and mythology. There are 88 official constellations which together cover the entire sky, each one of which refers to and delineates that particular region of the celestial sphere, the result being that every celestial object is described as being within one particular constellation or another.

DOUBLE STARS

Double stars are two stars which appear to be close together in space. Most of these are comprised of stars that are gravitationally linked and orbit each other, forming a genuine double-star system (also known as a binary star). Some double stars (known as optical doubles) are nothing more than chance alignments,

Mizar (left) and Alcor, the famous naked-eye double star in Ursa Major (see Ursa Major).

The galaxy UGC 12158 in Pegasus is classed as a barred spiral galaxy due to it having a central bar-shaped structure composed of stars.

and are made up of two stars that only happen to lie in the same line of sight as seen from Earth.

ECLIPTIC

Because the Earth is orbiting the Sun, the position of the Sun when seen against the background stars changes slightly from day to day, the overall effect of this being that the Sun appears to travel completely around the celestial sphere during the course of a year. This apparent path of the Sun is known as the ecliptic and is superimposed against a band of constellations we call the Zodiac (*see Zodiac*) through which the Sun appears to move.

GALAXY

A vast collection of stars, gas and dust bound together by gravity and measuring many light years across. Galaxies occur in a wide variety of shapes and sizes, including spiral, elliptical and irregular, and most are so far away that their light has taken many millions of years to reach us. Our solar system is situated in the Milky Way Galaxy, a spiral galaxy containing several billion stars. Located within the Local Group of Galaxies (see below), the Milky Way Galaxy is often referred to simply as the Galaxy.

INDEX CATALOGUE (IC)

References such as that for IC 2391 (in Vela) and IC 2602 (in Carina) are derived from their numbers in the Index Catalogue (IC), published in 1895 as the first of two supplements (the second was published in 1908) to his New General Catalogue of Nebulae and Clusters of Stars (NGC) by the Danish astronomer John Louis Emil Dreyer (1852–1926).

John Louis Emil Dreyer.

Between them, the two Index Catalogues contained details of an additional 5,386 objects (*see also* New General Catalogue).

LIGHT YEAR

To express distances to the stars and other galaxies in miles would involve numbers so huge that they would be unwieldy. Astronomers therefore use the term 'light year' as a unit of distance. A light year is the distance that a beam of light, travelling at around 186,000 miles (300,000 km) per second, would travel in a year and is equivalent to just under 6 trillion miles (10 trillion km).

LOCAL GROUP OF GALAXIES

A gravitationally-bound collection of galaxies known as the Local Group which contains over 50 individual members, one of which is our own Milky Way Galaxy. Other members include the Andromeda Galaxy (M31), the Large Magellanic Cloud, the Small Magellanic Cloud, the Triangulum Spiral Galaxy (M33) and many others.

Galaxies are usually found in groups or clusters. Apart from our own Local Group, many other groups of galaxies are known, typically containing anywhere up to 50 individual members. Even larger than the groups are clusters of galaxies which can contain hundreds or even thousands of individual galaxies. Groups and clusters of galaxies are found throughout the universe.

MAGNITUDES

The magnitude of a star is purely and simply a measurement of its brightness. In around 150 BC the Greek astronomer

GREEK ALPHABET

α Alpha	ε Epsilon	ι Iota	ν Nu	ρ Rho	φ Phi
β Beta	ζ Zeta	κ Kappa	ξ Xi	σ Sigma	χ Chi
γ Gamma	η Eta	λ Lambda	o Omicron	τ Tau	ψ Psi
δ Delta	θ Theta	μ Mu	π Pi	υ Upsilon	ω Omega

Hipparchus divided the stars up into six classes of brightness, the most prominent stars being ranked as first class and the faintest as sixth. This system classifies the stars and other celestial objects according to how bright they actually appear to the observer. In 1856 the English astronomer Norman Robert Pogson (1829–1891) refined the system devised by Hipparchus by classing a 1st magnitude star as being 100 times as bright as one of 6th magnitude, giving a difference between successive magnitudes of $\sqrt[5]{100}$ or 2.512. In other words, a star of magnitude 1.00 is 2.512 times as bright as one of magnitude 2.00, 6.31 (2.512 x 2.512) times as bright as a star of magnitude 3.00 and so on. The same basic system is used today, although modern telescopes enable us to determine values to within 0.01 of a magnitude or

Charles Messier.

better. Negative values are used for the brightest objects including the Sun (-26.8), Venus (-4.4 at its brightest) and Sirius (-1.46). Generally speaking, the faintest objects that can be seen with the naked eye under good viewing conditions are around 6th magnitude, whilst binoculars will allow you to see stars and other objects down to around 9th magnitude.

MESSIER CATALOGUE
References such as that for Messier 31 (in Andromeda) and Messier 57 (in Lyra) are derived from their numbers in the catalogue of deep-sky objects drawn up by the French astronomer Charles Messier (1730–1817) in the latter part of the 18th century. Deep-sky objects are objects

(other than individual stars) which lie beyond the confines of our solar system and include such things as star clusters, nebulae and galaxies.

MILKY WAY
The Milky Way is visible as a faint pearly band of light and is created by the combined glow of stars scattered along the plane of our Galaxy's disc, as seen from Earth. Given clear, dark skies, it is easily visible to the unaided eye and any form of optical aid will show that it is made up of many thousands of individual stars. Our solar system lies within the main plane of the Milky Way Galaxy and is located inside one of its spiral arms. The Milky Way is actually our view of the Galaxy, looking along the main galactic plane. The pearly glow we see is the combined light from many different stars and is visible as a continuous band of light stretching around the celestial sphere. Although the vast majority of these are too faint to be seen without optical aid, their combined light produces the glow that can be seen crossing the sky on dark, clear nights.

NEBULA
Nebulae are huge interstellar clouds of gas and dust. Observed in other galaxies as well as our own, their collective name is from the Latin 'nebula' meaning 'mist' or 'vapor', and there are three basic types.

Emission nebulae contain young, hot stars that emit copious amounts of ultra-violet radiation which reacts with the gas in the nebula causing the nebula to shine at visible wavelengths and with a reddish color characteristic of this type of nebula. In other words, emission nebulae emit

Norman Robert Pogson.

The Milky Way spans a beautiful starlit sky.

The Horsehead Nebula, a dark nebula in Orion, is beautifully silhouetted against the bright emission nebula IC 434.

their own light. A famous example is the Orion Nebula (M42) in the constellation Orion, which is visible as a shimmering patch of light a little to the south of the three stars forming the Belt of Orion.

The stars that exist in and around reflection nebulae are not hot enough to actually cause the nebula to give off its own light. Instead, the dust particles within them simply reflect the light from these stars. The stars in the Pleiades star cluster (M45) in Taurus are surrounded by reflection nebulosity. Photographs of the Pleiades cluster show the nebulosity as a blue haze, this being the characteristic color of reflection nebulae.

Dark nebulae are clouds of interstellar matter which contain no stars and whose dust particles simply blot out the light from objects beyond. They neither emit nor reflect light and appear as dark patches against the brighter backdrop of stars or nebulosity, taking on the appearance of regions devoid of stars. A good example is the Coal Sack in the constellation Crux, a huge blot of matter obscuring the star clouds of the southern Milky Way.

NEW GENERAL CATALOGUE (NGC)

References such as that for NGC 2129 (in Gemini) and NGC 6025 (in Triangulum Australe) are derived from their numbers in the New General Catalogue of Nebulae and Clusters of Stars (NGC), first published in 1888 by the Danish astronomer John Louis Emil Dreyer and which contains details of 7,840 star clusters, nebulae and galaxies (*see also* Index Catalogue).

The Helix Nebula (NGC 7293) in Aquarius.

PLANETARY NEBULA

Planetary nebulae consist of material ejected by a star during the latter stages of its evolution. The material thrown off forms a shell of gas surrounding the star whose newly-exposed surface is typically very hot. Planetary nebulae have nothing whatsoever to do with planets. They derive their name from the fact that, when seen through a telescope, some planetary nebulae look like luminous discs, resembling a gaseous planet such as Uranus or Neptune. Examples include the famous Ring Nebula (M57) in Lyra, the Ghost of Jupiter (NGC 3242) in Hydra and the Helix Nebula (NGC 7293) in Aquarius (*above*).

PRECESSION

The star Polaris in Ursa Minor currently marks the position of the north celestial pole. In other words, the Earth's axis is currently pointing towards Polaris, which means that if you were stood at the North Pole, Polaris would be located directly overhead. The daily rotation of our planet on its axis makes the rest of the stars in the sky appear to travel around Polaris, their

An illustration (not to scale) showing the Sun and major planets of the Solar System which, working outwards from the Sun, are Mercury, Venus, Earth, Mars, Jupiter, Saturn, Uranus and Neptune.

paths through the sky being centered on the Pole Star.

However, the position of the north celestial pole is slowly changing, this because of a 'wobble' in the Earth's axis of rotation. This wobble is known as

'precession' and is similar to that of a spinning top which is slowing down. Precession is caused by the combined gravitational influences of the Sun and Moon on our planet. Each resulting wobble of the Earth's axis takes nearly

26,000 years to complete, the net effect of precession being that, over this period, the north celestial pole traces out a large circle around the northern sky. For northern hemisphere observers, this results in a slow change in the apparent location of the north celestial pole. Polaris will eventually relinquish its position and Vega will become the Pole Star some 11,500 years from now.

SOLAR SYSTEM

The collective description given to the system dominated by the Sun and which includes the planets, minor planets, comets, planetary satellites and interplanetary debris that travel in orbits around our parent star.

STAR

A self-luminous object that shines through the release of energy produced by nuclear reactions at its core.

STAR COLORS

Stars are seen to have many different colors, a prominent example being the bright orange-red Arcturus in the constellation of Boötes, which contrasts sharply with the nearby brilliant white Spica in Virgo. Our own Sun is yellow, as is Capella in Auriga. Procyon, the brightest star in Canis Minor, also has a yellowish tint. To the west of Canis Minor is the constellation of Orion the Hunter, which boasts two of the most conspicuous stars in the whole sky; the bright red Betelgeuse and Rigel, the brilliant blue-white star that marks the Hunter's foot. Other stars with conspicuous colors include orange-red Aldebaran in Taurus

The whitish glow of the bright star Polaris in Ursa Minor, seen here surrounded by large complex dust structures known as molecular clouds.

The distinctive reddish glow of Betelgeuse at upper left contrasts with the white of Rigel at lower right in this image of Orion, which also depicts the Orion Molecular Cloud Complex, an expansive collection of bright nebulae, dark clouds and hot, young stars. The Orion Nebula M42 can be seen immediately to the south of (below) the three stars forming the Belt of Orion.

and Mu Cephei in the constellation of Cepheus. Mu Cephei, or the Garnet Star, is probably the reddest star visible to the naked eye in the northern skies, and binoculars will bring out the color very well.

The color of a star is a good guide to its temperature, the hottest stars being blue and blue-white with surface temperatures of 20,000 degrees K or more. Classed as a yellow dwarf, the Sun is a fairly average star with a temperature of around 6,000 degrees K. Red stars are much cooler still, with surface temperatures of only a few thousand degrees K. Betelgeuse in Orion and Antares in Scorpius are both red giant stars that fall into this category.

When we look up into the night sky the stars look much the same. Some stars appear brighter than others but they all look white. If the stars are looked at through a pair of binoculars some appear to be different colors. Many stars look quite orange in color and some have a blue/white tint. The color of a star depends on its surface temperature and indirectly on the size.

STAR NAMES

Over 200 stars have proper names, usually of Roman, Greek or Arabic origin. However, only a couple of dozen or so are used regularly, examples of which include Arcturus in Boötes, Vega in Lyra, Deneb in Cygnus, Spica in Virgo and Betelgeuse in Orion.

The system whereby Greek letters are assigned to stars was introduced by the German astronomer Johann Bayer (1572–1625) in his star atlas *Uranometria*,

WHAT STAR?

There are only 24 letters in the Greek alphabet, which means that the fainter naked-eye stars need an alternative system of classification. The system in popular use is that devised by the English Astronomer Royal John Flamsteed (1646–1719) in which the stars in each constellation are listed numerically in order from west to east. Although many of the brighter stars have both Greek letters and Flamsteed numbers, the latter are generally used only when a star does not have a Greek letter, as is the case with

1 Geminorum (*see Gemini*) or 4, 5, 6 and 10 Serpentis (*see Serpens*).

STAR CLUSTERS

Although most of the stars that we see in the night sky are scattered randomly throughout the spiral arms of the Galaxy, many are found to be concentrated in relatively compact groups, referred to by astronomers as star clusters. There are two main types of star cluster – open and globular. Open clusters, also known as galactic clusters, are found within the

John Flamsteed, the first Astronomer Royal of England.

published in 1603. Bayer's system is applied to the brighter stars within any particular constellation, which are given a letter from the Greek alphabet, followed by the genitive case of the constellation in which the star is located. This genitive case is simply the Latin form meaning 'of' the constellation. Examples are the stars Alpha (α) Boötis and Beta (β) Centauri, which translate literally as 'Alpha of Boötes' and 'Beta of the Centaur'.

As a rule, the brightest star in a constellation is labeled Alpha (α), the second brightest Beta (β), the third brightest Gamma (γ) and so on, although there are some constellations where the system falls down. An example is Gemini, where the principal star (Pollux) is designated Beta (β) Geminorum, the second brightest (Castor) being known as Alpha (α) Geminorum.

The Great Hercules Cluster M13 (NGC 6205), one of the finest globular clusters in the sky.

Rich star fields surround the magnificent Sword Handle Double Cluster in Perseus (see Perseus).

globular clusters, there are three famous examples which can be spotted with the naked eye. These are the Great Hercules Cluster (M13) in the constellation Hercules, Omega Centauri in Centaurus and 47 Tucanae in Tucana.

VARIABLE STARS

A variable star is a star whose brightness varies over a period of time. There are many different types of variable star, although the variations in brightness are basically due either to changes taking place within the star itself, such as L2 in Puppis or Mira in Cetus, or the periodic obscuration, or eclipsing, of one member of a binary star by its companion, such as Algol in Perseus.

ZENITH

The point in the sky directly above the observer.

ZODIAC

The band of 12 constellations which straddles the ecliptic through which the Sun appears to travel throughout the course of a year. The constellations which form the Zodiac are Aries, Taurus, Gemini, Cancer, Leo, Virgo, Libra, Scorpius, Sagittarius, Capricornus, Aquarius and Pisces.

main disc of the Galaxy and have no particularly well-defined shape. Usually made up of young hot stars, over a thousand open clusters are known, their diameters generally being no more than a few tens of light years. They are believed to have formed from vast interstellar gas and dust clouds within our Galaxy and indeed occupy the same regions of the Galaxy as the nebulae. A number of open clusters are visible to the naked eye, including Praesepe (M44) in Cancer and perhaps the most famous open cluster of all, the Pleiades (M45) in Taurus.

Globular clusters, as their name suggests, are huge spherical collections of stars located in the area of space surrounding the Galaxy. With diameters of anything up to several hundred light years globular clusters typically contain tens of thousands of old stars and little or none of the nebulosity seen in open clusters. When seen through a small telescope or binoculars, they take on the appearance of faint, misty balls of grayish light superimposed against the background sky. Although some form of optical aid is usually needed to see

OPPOSITE: The Zodiac is formed from the 12 constellations depicted here.

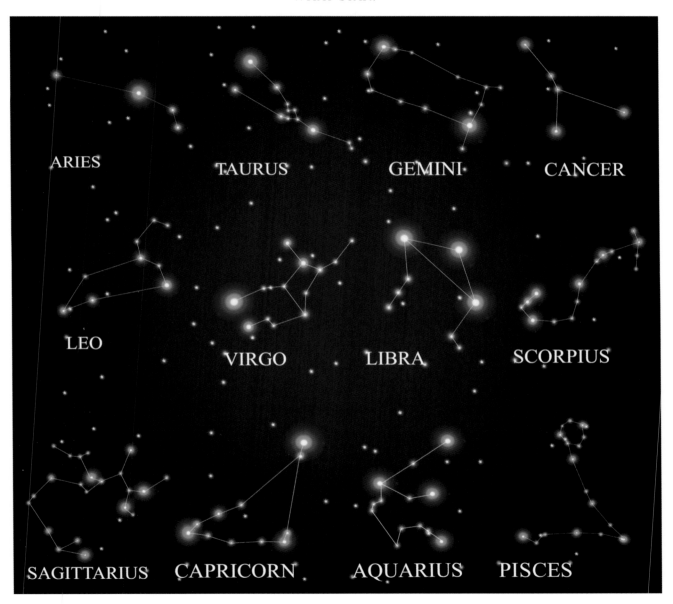

INDEX

ACKNOWLEDGEMENTS

All star charts are supplied by Gary Blackmore.

NASA/Hubble Space Telescope/European Space Agency.
Pages: 37 below, 41, 43, 45 both, 47, 52, 61, 71 both, 64, 81, 84, 88, 119, 122, 125, 141, 142, 147, 157, 158, 159, 169, 172 left, 184, 190, 191 above, 192, 194, 201 both, 202 left, 211, 214, 217, 218, 219, 227 both, 230, 238, 240.

European Southern Observatory.
Pages: 37 above, 67, 74, 75, 94, 95, 103 below, 120, 149, 153, 155, 167, 181 above, 196, 233, 236.

Ray Emery
Pages: 59 above, 237.

Kevin Taylor
Pages: 34, 49 left.

@Shutterstock.com
2, 3, 4, 8-9, 10 all, 11, 12, 13, 14-15, 16, 18, 19, 20, 22, 24, 28, 30-31, 33, 35, 49 (right) 56, 57, 59 below, 62, 68, 71 below, 72, 77 below, 78, 79, 81 below, 82, 87, 90, 98 both, 99, 101, 103 above, 105, 110, 111, 114, 115, 118, 123, 127, 130, 133, 139, 146, 148, 151, 162, 163, 164, 165, 172 right, 173, 178, 181 below, 187, 188, 189, 195 above, 197, 199, 202 right, 203, 204, 208 right, 209, 213, 221, 222, 229, 235, 239, 243, 244, 245, 246, 247, 249, 25, 251.

Wikimedia Commons
Page 248 Rogelio Bernal Andreo, page 113 David Bishop, page 126 Egres73, page 134 San Estaban, page 72 Francesco Malafarina, Pages 137, 152, 215 Roberto Mura, pages 98 below, 145, 195 Ole Nielsen, page 208 TVance, page 54 Wayne Young, page 191 below Hunter Wilson,